The

Mycocultural Revolution

TRANSFORMING OUR WORLD WITH MUSHROOMS, LICHENS, AND OTHER FUNGI

Peter McCoy

FOREWORD BY ROBERT ROGERS, RH (AHG)

Microcosm Publishing
PORTLAND, ORE

THE MYCOCULTURAL REVOLUTION
Transforming Our World with Mushrooms, Lichens, and Other Fungi
© Peter McCoy 2022

ISBN 978-1-62106-514-2
This is Microcosm #289
First Published November, 2022
This edition © Microcosm Publishing, 2022

For a catalog, write or visit:
Microcosm Publishing
2752 N Williams Ave.
Portland, OR 97227
www.Microcosm.Pub/Myco

To join the ranks of high-class stores that feature Microcosm titles, talk to your rep: In the U.S. **COMO** (Atlantic), **ABRAHAM** (Midwest), **BOB BARNETT** (Texas, Oklahoma, Arkansas, Louisiana), **IMPRINT** (Pacific), **TURNAROUND** (UK), **UTP/MANDA** (Canada), **NEWSOUTH** (Australia/New Zealand), **Observatoire** (Africa, Middle East, Europe), **Yvonne Chau** (Southeast Asia), **HarperCollins** (India), **Everest/B.K. Agency** (China), **Tim Burland** (Japan/Korea), and **FAIRE** and **EMERALD** in the gift trade.

Disclaimer

Global labor conditions are bad, and our roots in industrial Cleveland in the 70s and 80s made us appreciate the need to treat workers right. Therefore, our books are MADE IN THE USA

Library of Congress Cataloging-in-Publication Data

Names: McCoy, Peter (Mycologist), author.
Title: The Mycocultural revolution : transforming our world with mushrooms, lichens, and other fungi / Peter McCoy
Other titles: Transforming our world with mushrooms, lichens, and other fungi
Description: Portland : Microcosm Publishing, [2022] | Summary: "Discover the glorious world of mushrooms, lichens, and micro fungi, as described by Peter McCoy, one of today's foremost experts in the field. Covering the essential information and skills for identifying, cultivating, and celebrating the uniqueness of fungi, this book enables anyone to quickly and easily engage in the art and science of mycology-the study of fungi. Mycology offers vast opportunities to enhance our lives, support our communities, and heal the environment. This first-of-its-kind introductory text is accessible for anyone just getting started in mycology, as well as for those seeking a fresh perspective on this important science. Learn general mycological facts, essential information and skills for identifying common mushroom types, foraging tips, delicious recipes, a growing guide, mycoremediation (using fungi to treat contaminated areas in our environment), mushroom-based crafts, and so much more!"-- Provided by publisher.
Identifiers: LCCN 2021052051 | ISBN 9781621065142 (board)
Subjects: LCSH: Mycology--Handbooks, manuals, etc. | Mushrooms--Handbooks, manuals, etc. | Fungi--Handbooks, manuals, etc.
Classification: LCC QK603 .M388 2022 | DDC 579.5--dc23/eng/20211108
LC record available at https://lccn.loc.gov/2021052051

MICROCOSM · PUBLISHING

Microcosm Publishing is Portland's most diversified publishing house and distributor with a focus on the colorful, authentic, and empowering. Our books and zines have put your power in your hands since 1996, equipping readers to make positive changes in their lives and in the world around them. Microcosm emphasizes skill-building, showing hidden histories, and fostering creativity through challenging conventional publishing wisdom with books and bookettes about DIY skills, food, bicycling, gender, self care, and social justice. What was once a distro and record label started by Joe Biel in a drafty bedroom, was determined to be *Publisher's Weekly's* fastest growing publisher of 2022 and has become among the oldest independent publishing houses in Portland, OR and Cleveland, OH. We are a politically moderate, centrist publisher in a world that has inched to the right for the past 80 years.

Did you know that you can buy our books directly from us at sliding scale rates? Support a small, independent publisher and pay less than Amazon's price at www.

MICROCOSM.PUB

Dedicated to anyone learning from the past to heal the present and spawn a brighter future.

CONTENTS

FOREWORD

*T*first met Peter on my second trip to Telluride in 2014. I was invited to give a few presentations on medicinal mushrooms, and the organizers lodged me in a house with a few other speakers. Peter had grabbed a secluded, second floor bedroom, and I barely saw him for the entire five days, as he worked fervently on his excellent, and highly successful, *Radical Mycology: A Treatise on Seeing & Working with Fungi* manuscript.

A few years later, Peter was invited to present at the Alberta Mycological Association's Annual Dinner. I had served as vice-president for many years, and I quickly agreed to greet him at the airport.

The next evening, Peter gave a thoughtful, passionate presentation on the importance of mushrooms to all our lives, mixed in with his vision of community-based citizen science and the future of fungi.

And thus, I was very honored when Peter invited me to share a few words about his latest book.

It is said, in the publishing world, that a hugely successful first book can lead to an author's concerns regarding their sophomore publication.

There are no worries here! Peter has expanded on his futuristic vision of the importance and versatility of fungi, shining a light on their significant contributions.

The Mycocultural Revolution continues an important dialogue around the mushroom's numerous benefits today and spawning forward.

The literary journey begins with a history of his involvement in the radical mycology movement (including the original zine), the 2016 publication of *Radical Mycology*, and some background around the five previous radical mycology convergences, as well as numerous event tours around the continent.

The Mycocultural Revolution covers lots of new, fertile ground, including great tips on wild foraging, and the history and benefits of fungal fermentation, including detailed recipes for various myco-cultures, including tempeh, mead, miso, and the delicious corn smut (huitlacoche).

The book touches on the better-known medicinal mushrooms and how anyone can produce their own safe and efficient dual extraction tinctures.

Cultivation techniques, covered extensively in his previous book, are reviewed and refined, with an interesting integration of fungi and gardening.

I have been recording mushroom music for many years, so I was thrilled to see this concept mentioned, along with unique philosophical ideas around creative mushroom movements, dance, and expressions, all in the context of cultural projects. So fascinating!

Fungal pigments and dyes are mentioned, along with the yet to be fully explored importance of myco-remediation, and an examination of fungal endophyte relationships within the plant kingdom. As an herbalist for a half century, I am increasingly convinced the health and medicinal properties of herbs are due, in large part, to this fungal relationship.

Above all, Peter's passion for the fungal kingdom shines through, personified by his love and commitment to helping create a healthy, sustainable, and vibrant planet and people. May the spores be with you!

—*Robert Rogers, RH (AHG)*
May 2022 Edmonton, Alberta

PREFACE

*M*ycology, the study of fungi, is one of the most inspiring, mysterious, exciting, and empowering topics I have ever studied. And yet, like almost everyone I've met, I was taught next to nothing about mushrooms, molds, or yeasts growing up. Of what little I learned from my peers and the media's depiction of fungi, I knew that most species were to be feared and avoided—save for the cold, tasteless species found at diner salad bars or, for the devious, the few mushrooms that could alter your perception of reality. Beyond that limited presentation, fungi were almost never mentioned in any context, and so I unconsciously assumed that they didn't matter.

It was only by luck that my interest in these incredible organisms first developed when, at the age of 15, my older brother suggested I try my hand at growing edible mushrooms. I had tended a vegetable and herb garden in our family's backyard for years, so learning to grow a new crop was more intriguing than intimidating to me. But as I didn't have a strong relationship with eating mushrooms, I wasn't that interested in growing fungi for food. Rather, my interest in eclectic topics cultivated

If we are only told by our teachers and the media that fungi will either hurt us or make us trip, our collective perspective on mycology will stay just as limited. Fungi are so much more than what we have been told to think about them.

a feeling in me that I couldn't ignore: the alluring oddity of a human-fungal relation was a mystery worth exploring, even if I was unsure of where it might lead.

After borrowing the few books on mushroom cultivation available at my local library—all of which were written for the large-scale fungi farming industry—I began the first of many attempts at cultivating gourmet mushrooms in my bedroom. Using what little engineering skills I had at the time, I built humble versions of the elaborate equipment described in the manuals I was reading, with a poorly sealed "still air box" (used for making clean transfers of mushroom tissue) and a less-than-optimal fruiting chamber (which mimics the natural environment in which mushrooms grow) being my clunky results. I grew more mold than mushrooms in those early years. Luckily, though, I never got discouraged by my failures, but rather saw each as well-earned steps on the path toward a greater understanding of the many ways of fungi.

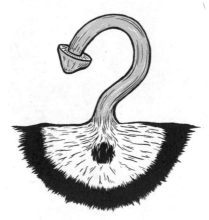

The myco-illiteracy of the world will quickly be overcome when more people start asking where fungi are missing from analyses, and where they've been overlooked historically.

As I continued to experiment with mushrooms throughout my teens, I also developed a strong interest in other beneficial aspects of mycology, such as its *many* influences on plants, animals, and ecosystems. Eventually, my understanding of the importance of fungi's influence on all life worked its way into my analysis of most global issues—a web of relationships that only increased in size and complexity as I got older. By my late teens, I often found myself wondering how mycology influenced politics, economics, and even how humans relate with each other and the environment. I knew that fungi were critical to plant and soil health, and that mushroom growing could reduce food shortages and, potentially, resolve health crises. And yet, all of the books and films I used to learn about these broader topics *never* incorporated a fungi-informed analysis. A systemic problem was presenting itself to me: not only did my peers lack consideration for fungi, but so did experts.

At the age of twenty, while volunteering with several activist organizations in New York City, my awareness of the West's mycological blind spot expanded to edges of society where complex social and environmental issues were often at the fore of discussion. One group, the New York Freegans, that I often worked with focused on addressing New York City's food waste by leading nightly "trash tours," during which we would take individuals and media outlets from around the world to the sidewalks facing several of lower Manhattan's grocery stores. Here, we would open the shops' trash bags and show just a small percentage of the thousands of pounds of fresh food that goes to waste in the city each day. During interviews with the media, we would explain the social and ecological impacts of food waste and consumerism, as well as how salvaging some of society's discarded excess can not only offset its ecological burden, but also serve as a reminder that food is a basic human right and something that should be shared—especially when a surplus is destined for the dumpster.

Working with the New York Freegans (www.freegan. info), I learned how to engage with the media and articulate challenging concepts. More importantly, I also learned how to discuss uncommon topics and present alternative paradigms that test popular assumptions—invaluable lessons that I carry with me and draw on to this day.

Though my fellow Freegans spent much of their time working to make the world a bit better, the impact that fungi had on our causes was rarely discussed during our tours or private meetings. When I would attempt to describe the relationships I saw between fungi, environmental health, food production, waste recycling, job creation, and even designing decentralized social structures (that often mimic the growth habits of fungi), my peers were rarely able to match my enthusiasm as the intersections I suggested were not part of the activist milieu at the time. After several efforts to convey what I felt, but could only articulate in limited terms, I recognized the need to hone my arguments and increase my understanding of the connections I sensed—all in the hope of advancing dialogues of liberation with my peers.

This process was an internal and silent one that only resurfaced when I started college in Olympia, Washington and met others who shared my combined interests in mycology, activism,

Radical Mycology zine assembly headquarters in Olympia, Washington (ca. 2006)

and culture creation. Thanks to their excitement around my ideas, I eventually decided to put my thoughts to paper in a zine titled *Radical Mycology*—a double entendre expressing both the text's eclectic approach to an eclectic topic, as well as a shout out to the politics-laden subcultures that had fostered my own philosophies.

Initially, I assumed that no more than a few dozen people might read *Radical Mycology*, only to find after a few short years that I had sold thousands of handmade copies to individuals, zine distros, punk houses, and bookstores around the world. Inspired by the positive feedback that the zine had so quickly received, in 2011 I worked with four like-minded friends to organize the Radical Mycology Convergence (RMC): a weekend-long gathering focused on building a broad community around the skills and sentiments expressed in my zine.

Soon after announcing the event, we realized that the RMC's emphasis on a grassroots form of mycological activism was unprecedented and greatly sought after by the hundreds of fungi-lovers who attended the event from around the world. Many attendees told me throughout the weekend that they had never before met anyone who shared their interest in DIY mycology and were now surrounded by dozens of new, like-minded friends. By the end of the RMC, dozens of people told me that the event had changed their life, gave them newfound hope, and helped them feel less alone in their long-time—and often secret—love for fungi. At the same time, several elder mycologists told me that the Convergence inspired excitement in them for the future of fungi, which they had long feared was destined for the pitfalls of commercialization and capitalism.

Reflecting on the RMC after it ended, I realized that not only was there a significant lack of applied mycology teachers in the world (i.e. those who teach others how to grow fungi for food or medicine, or otherwise integrate fungi into practical applications), but also that the skills and philosophy that the Radical Mycology ethos offered was greatly needed by others. I decided then that the best service I could offer current and future generations was to further elaborate on the ideas behind the zine and the RMC and to teach more people about all that working with fungi offers.

Over the coming years, I helped organize several more RMCs and lead several workshop tours around North America, until finally sitting down in 2014 to begin writing my first book, *Radical Mycology: A Treatise on Seeing and Working With Fungi* (Chthaeus Press, 2016). The text distilled the information, skills, and insights I had acquired over the previous 15 years—all in the hopes of creating the kind of reference guide to mycology that I had always wanted growing up. Following the book's positive reception, in 2017 I founded MYCOLOGOS (www.mycologos. world), an online mycology school and organic mushroom farm based in Portland, Oregon, with the intention to create an ever-

Radical Mycology workshop at Interference Archive in Brooklyn, New York (ca. 2014)

more accessible platform for people around the world to learn mycology.

There have been five Radical Mycology Convergences (RMCs) held around the U.S. over the last decade, as well as several North American Radical Mycology workshop and event tours—each of which worked closely with food, environmental, and social justice organizations to increase access to mycology for people of all backgrounds. Attendees of the RMCs have even gone on to host their own events inspired by the Convergence and to evolve dialogues around how best to work with fungi in their local community.

Just as at the first RMC, Radical Mycology's mission remains to celebrate and advocate for all fungi through hosting unique gatherings, curating media, and by providing support to fungi-centric artists—with new ideas and projects coming to the fore each year. You can learn more about our work at www.radicalmycology.com.

HOW TO READ THIS BOOK

This book is the latest step in my journey with fungi. Whereas the zine *Radical Mycology* presents the foundational concepts of a philosophy and social movement that has been growing for over a decade, and the book *Radical Mycology: A Treatise on Seeing and Working With Fungi* serves as an extensive reference guide to all things related to fungi, *The Mycocultural Revolution* is a summary of the first steps I suggest for anyone wanting to engage in the modern mycological era.

The book's introduction lays the foundation for the text's three main sections by first covering a brief history of fungi in human culture and the important changes occurring in the world of mycology today. The introduction concludes with an explanation of how fungi are biologically and ecologically unique, and why these details are essential for engaging with the skills in the rest of the text.

In Part I, we begin looking at the ways that you can bring fungi into your daily life by discussing three of the most popular applied mycology skills. Chapter 1 demystifies the art of mushroom and lichen foraging, enabling beginners nearly anywhere in the world to safely and confidently get started in the practice. Chapter 2 covers several ways to make nourishing foods and drinks with fermenting fungi, as well as some of my favorite recipes for cooking mushrooms—all of which are sure to help convert almost anyone who dislikes eating mushrooms into a devout fungi-lover. And Chapter 3 introduces the healing benefits of medicinal mushrooms by providing several ways to incorporate their potent effects into your daily practice, while also touching on the potential health benefits of psychoactive fungi.

In Part II, we expand our work with fungi to skills that support our friends, family, and greater community. Chapter 4 introduces practical skills for low-cost, low-waste, and high-yield mushroom cultivation that can be applied in nearly any setting, even with limited resources. And Chapter 5 explores several ways to build a cultural language around the beauty of mycology through several fungi-infused art projects.

In Part III, we conclude by looking at how the skills and insights of Part II can readily extend to the communities and lands around us. In Chapter 6, we first consider several easy ways to integrate fungi into food production and gardening systems, while in Chapter 7 we cover ways by which fungi can offset pollution, as well as how anyone can help advance the science of mycology on the whole. To close out the book, the most helpful mushroom species to know about are profiled in depth and several appendices are provided to help you dive deeper into all that we cover.

To make such a large topic as mycology feel approachable for everyone, I have kept technical language and the finer-grain details on most of these topics to a minimum, and so suggest that readers seeking more information on a given topic refer to *Radical Mycology: A Treatise on Seeing and Working With Fungi*. If you are new to mycology, reading the chapters in their presented order is suggested as each topic builds upon the last. That said, feel free to skip around as this book is primarily meant to serve as a how-to starter guide for years of walking your own path of fungi-filled discovery.

Working with fungi and learning to see them a bit clearer over the last twenty years has transformed my understanding of the world and of myself in more positive ways than I can count. It has been a winding, humbling, inspiring, and surprising road—in *many* more ways than I ever could have imagined when I took my first steps into its uncharted terrain.

I am grateful to have had the opportunity to engage in this work for so long and am thankful to the good folks at Microcosm Publishing for providing me with the space to share some of the insights I have gleaned along the way in these pages. I am also thankful for the time you lend these topics, whether to support your life or your land base. It is my hope that in sharing my latest joys in the world of fungi with you, that I might also impart new ways of seeing and being in the world we share. In return, I hope to someday learn from all that mycology has brought to you, whether at the next Radical Mycology Convergence or elsewhere. Fungi need as many advocates and ambassadors as will embrace them. So, what are you waiting for? Join us!

For the Fungi!
Peter McCoy
Portland, Oregon
Mushroom Season, 2022

INTRODUCTION: MYCOCULTURE RISING

*F*or over two decades, a global uprising of interest in mushrooms has brought the notion that fungi matter, shifting it from its centuries-old refuge on the fringes of society towards the center of nearly every aspect of modern life. No longer are mushrooms as feared as they were just a few years ago. Society has increasingly embraced them for their curious ways and seemingly endless ability to help solve some of the world's most challenging issues. People around the world are rapidly discovering that the simple acts of growing gourmet and medicinal mushrooms could reduce global hunger, alleviate some of the most challenging diseases of the 21st century, improve topsoil health, and even reduce pollution. Indeed, it seems that with each new research paper published on fungi, that the science of mycology only grows in its ability to spawn a better world through working with these incredible organisms.

Reflecting on mycology's history, this cultural shift is arguably due in large part to advancements in technology, which have not only refined our understanding of fungi but also increased public access (thanks, internet!) to mycological knowledge that had long been secluded to a small number of universities. As we now enter this latest stage in the longstanding lineage of human-fungal relations, we can reflect to find that, like many aspects of the human story, the human-fungal relationship has progressed for countless generations. Though lesser-told than our tales of developing corn from teosinte or of domesticating the horse, the fungi-infused aspects of our ancestry are integral threads that have woven throughout our growth over the eons.

Such a long saga can be most readily appreciated by outlining its major stages, which I see as divided into four epochs, or volumes, in the story shared by humans and fungi. The first of these is a millennia-long compendium of fungi-infused myths, artworks, and customs developed by traditional peoples around the world. Across Eurasia, Africa, and the Americas, we find rich expressions of humans working with molds, mushrooms, lichens, and yeasts as medicine, food, and tools for ritual—with each practice relaying new means for perceiving and respecting the unique traits of the local's most revered species. This volume of our story is the longest of the four, and yet is still being revised as new insights into the lifeways of the past reveal ever-more complex historical relationships between our human and fungal ancestors.

This early era gave way to the second volume of our story in the 1750s, when mycology transitioned from a cultural inheritance to a formalized study. Under the honing blade of the scientific method, the following two centuries saw improvements in our ability to describe and cultivate fungi, as well as a reframing of their value in terms that often weighed economic relevance over ecological significance or cultural preservation. By the early twentieth century, complex methods for exploiting the compounds produced by molds and yeasts had become commonplace across industries, just as commercial mushroom growing had become a mechanized and artificial version of an art that had long been based in mimicking natural processes. The result of this divorce of man from fungi created a widespread belief that mycology on the whole was technologically inaccessible to the average person, and therefore irrelevant for discussing—let alone promoting—in popular culture.

In the 1950s, the narrative was once again rewritten when psychoactive mushrooms were reintroduced into Western culture. In the decades that followed, subcultural interest in illicit mushrooms spread alongside advancements in understanding fungal biology and ecology. By the end of the 20th century, mycologists came to recognize the profoundly important influences of fungi on plants and animals to greater degrees than ever before—with the conclusion being that fungi are critical for the sustainment of life on Earth.

Alongside this research, this period also saw industrial mushroom cultivation methods of prior decades simplified for home-scale approaches as a growing number of cultivators and researchers realized that the practice had applications beyond food production. By the start of the twenty-first century, mushroom growing had become so easy it could be practiced by nearly anyone, whether in a makeshift bedroom laboratory, or in the kitchen while making dinner. As the efforts of fungal researchers and mushroom cultivators increasingly overlapped, the internet and social media helped propel the democratization of mycology, finally bringing the many benefits of the science to people of all backgrounds and, ultimately, creating the greatest groundswell of *mycophilia*, or love for fungi, that the world has ever known.

MURMURS OF A REVOLUTION

The recent revolutionary changes in the mycological landscape mark the start of our current era in the human-fungi story—a time calling for reassessment across many facets of society and fundamental shifts in the status quo's standards. During this

latest era of working with molds and mushrooms, we find the chance to revisit long-standing assumptions about what it means to engage with any fungal species and to increasingly ask how these organisms have been overlooked historically and where they are still disregarded. As time goes on, the answers to these questions will soon lead to mycology becoming more fully integrated into human activity, just as taboos around mushrooms are increasingly being replaced by a fascination for their unknown potential. Whereas just a few years ago leading mycologists expressed concern that the study of fungi was at risk of fading from academia (where it still struggles to gain the attention and funding it deserves), the revival of mycology as a people's science has instilled newfound hope in all who have long promoted the powers of fungi that might heal the planet, if only given time and space.

As our current volume in this story begins to be written, both novice and experienced mycorevolutionaries have an opportunity to set the tone for coming chapters. The rise in interest in mycology will require more practitioners to join the movement and provide fungi-centric skills to their community. An influx of fresh perspectives will help refine still burgeoning areas of research, such as low-waste mushroom cultivation and the design of sustainable cities that account for the many applications and roles of fungi in our lives. Being one of the youngest, largest, and least understood fields of study, mycology offers a rare opportunity for experimentation.

Changes to our cultural relationship with fungi—our *mycoculture*—also transform the individuals who co-create the revolution. Learning to see the mystery and magic of fungi in all of their forms dramatically changes how one perceives the natural world and the ways by which it is impacted by humans. Fungi are everywhere around us. Endophytic fungi fill the tissues of every

The history of human-fungal relations is recorded across four volumes, with the latest only just beginning to be written. What stories will you add to this legacy?

plant and mycorrhizal fungi build the life-sustaining soil beneath our feet. In doing so, they shape the growth of every habitat we enter and guide the evolution of all life in ways we've only started to comprehend.

And yet, because fungi are largely absent from popular education and mass media, most people in the West are never offered the chance to view the world through a mycological lens, but instead inherit a narrow and often fear-based perspective on the value of fungi—a near-global unawareness that leads to avoiding these organisms and what they represent. In the mycocultural revolution, unlearning these negative perceptions is the first act of liberation that we all must take in reclaiming our right to engage with wild fungi and unite their many gifts with the stories of our lives.

Creating potent mushroom-based medicines at home and cultivating fungal gardens are just some of the physical acts of empowerment that mycology provides. In learning these and other fungi-centered skills, we're given the opportunity to tap into forgotten traditions, to be inspired by the lessons expressed in fungal ecology, and to develop personalized ways of engaging with these incredible organisms—the myco-rituals of daily practice that can never be mediated or purchased. Such personal customs represent small but critical additions to the ongoing story of human-fungal relations, which, when combined with those crafted by others, will pave a shared path toward the most positive and potent future of fungi.

TRAINING AND TACTICS

Adding to the mycoculture is a rare opportunity to guide a major shift in human culture—a gift that will be best served by recognizing its precious nature and taking thoughtful steps forward as more people join the mycological march. Whether the future of our story will be centered on the profits of entrepreneurs or on connections to the natural world will be determined by the choices we make during the current era.

To ensure that this discussion is accessible to all who wish to take part, we must increase familiarity with the core concepts and skills of traditional and applied mycology. The world's myco illiteracy must be dissolved to ensure that the coming advancements in mycology will be easy to understand and accepted by the majority of people. This book's coming chapters provide the skills and concepts1 that I believe are the essential first steps for shaping such a healthy narrative. But, before we can jump into those bigger topics, the basics must be covered—with the first being a clarification of just what fungi are and what makes them so special.

ORGANIZING CHAOS

A number of the traits that separate mushrooms and molds from plants and animals can be described using terms that are analogous to plant and animal biology. But many of the most important and unique fungal qualities are just that: *fungal things that only fungi do*. In classical taxonomy (the science of naming and categorizing life on Earth), fungi are first distinguished from other organisms by how they obtain nutrition and how their cells are structured. Unlike the small and simple prokaryotic cells of bacteria, the eukaryotic cells of fungi, plants, and animals are large and complex. Under a microscope, animal cells look like tiny water balloons made of an outer layer, the cell membrane, which contains several different structures, or organelles, inside. The

A simplified representation of the world's biodiversity, with smaller celled microbes on the left and larger celled eukaryotic organisms on the right. Being the decedents of fungi, animals and plants branch off of the mycelium that forms the backbone of this web of life. As a symbiotic community of organisms, lichens are not connected to the web, but are considered special cases that live in their own world of emergent splendor.

cells of plants and fungi also have a membrane and organelles, but also an additional, rigid layer beyond the cell membrane called the cell wall. This strong layer enables plants to grow tall without having an internal skeleton, and it gives fungi the ability to grow in deep soil and dense wood without being crushed.

A major difference between fungi and plants is in the compounds that comprise their cell walls. In plants, cellulose, hemicellulose, pectin, and lignin are the main wall components, whereas fungal cell walls are made up of large (and often medicinal) sugars, and chitin, the stiff compound that makes up lobster shells and insect exoskeletons. Additionally, fungi have

PROKARYOTES VS. EUKARYOTES

The prokaryotic cells of bacteria are smaller and simpler than those of eukaryotic organisms (i.e. animals, plants, and fungi). Whereas eukaryotes contain their DNA inside of a nucleus, the prokaryotic DNA floats freely. A primary difference between animal, plant, and fungal cells is in their outer layer design. Animal cells are contained by a non-rigid cell membrane, while plant and fungal cells host an additional cell wall that provides rigidity. Plant cell walls contain cellulose, hemicellulose, pectin, and lignin, while fungal cell walls are composed of chitin and high-weight polysaccharides.

a unique compound in their cell wall called ergosterol, which is similar to the cholesterol in our bodies. Fungi also store their extra energy in a chemical known as trehalose and in several types of alcohols, while plants store their extra energy as starch and animals store it as glycogen or body fat.

Where things get really interesting, though, is when one fungal cell becomes many. The majority of fungi are multicellular, just as most plants and animals are. Unlike plants and animals, however, fungi do not form a variety of tissues or organ types. Instead, ninety-nine percent of fungal species—from the tiniest molds to the largest mushroom cultures in the world—are composed entirely by one type of tissue: the mighty and magnificent mycelium.

The features and functions of mycelium—along with the insights that come from observing them—underscore the rest of this book and, really, much of mycology, as it is mycelium that provides most of the important fungal functions in the environment and is fundamentally what most fungi are comprised of. And yet, mycelial biology (i.e. how this tissue grows and functions) remains a rather mysterious topic, with many of its details yet to be fully understood. For the purposes of this book, the essential concepts to appreciate about the primary fungal tissue are:

- Mycelium is a network of one-cell-thick threads known as hyphae. Each hypha in this network grows, or, rather, extends, at its tip (that is, the side walls do not extend once that are set in place).

- Inside of each hyphal tip, a unique-to-fungi organelle known as the Spitzenkörper acts like a microscopic brain that interprets and responds to environmental signals, such as the presence of food or environmental competitors (e.g. viruses, bacteria, and other fungi).

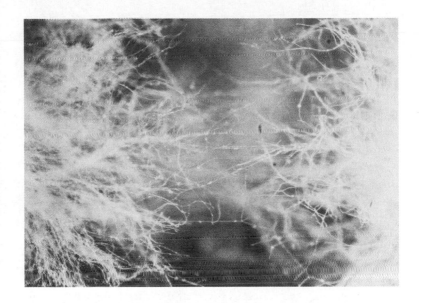

- As a part of its response to the environment, the Spitzenkörper releases a wide array of compounds outside of itself to surround the hypha. Some of these compounds break down food in the fungus's path, while others ward off competitors trying to eat the fungus or its food. Each hypha lives in this bath of digestive and defensive exometabolites as it grows through its substrate, or food. This external digestion is another fungal feature that is not found in plants (which mostly photosynthesize to obtain energy) nor animals (which digest their food via an internal stomach).

- Combined, these functions make each hyphal tip the mycological "business end" of the mycelium—with the network's leading (or radical) edge constantly adapting to new circumstances and challenges to ensure the survival of the entire culture.

To reproduce, most fungi release microscopic spores into the environment. There are many types of spores and many ways by which fungi produce them, making the stages of fungal sex surprisingly difficult to generalize. For our purposes, a summarized version of a Shiitake mushroom's life cycle conveys the most important points of this otherwise complex topic. In Shiitake mushrooms, we find spores produced under the mushroom's cap on the surface of gills from which they eventually eject into the air to travel on a breeze or the fur or feathers of a passing animal, or, most commonly, to land just a few feet away from the mushroom.

If a spore lands in an area where food and water are present, it will germinate, sending out an initial hypha that quickly branches to form a mycelial network. Most spores have half of the genetics of their parent mushroom, and so will seek out a sexually compatible mate to fuse, or anastomose, with and thereafter create a genetically complete mycelial network that will later be

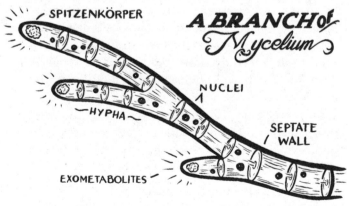

A single branch of mighty mycelium is a microcosm of fungal magnificence. Constantly branching as it navigates the world, each hyphal thread is helmed by a Spitzenkörper that surveys the environment and controls the release of protective and digestive exometabolites. Cross-walls host an opening (a septa) that controls the flow of cellular contents, including nuclei, throughout the network.

able to produce its own spore-bearing mushrooms that will, in turn, repeat the life cycle for another generation.

Other types of fungi produce spores on the surface of their mycelial mat (e.g. many of the molds that grow on old food), while others, such as some soil-dwelling species, convert a single cell at the end of a hypha into a spore that soon thereafter falls off of the mycelial network. This variety in reproduction strategies is part of what lends to the resilient nature of fungi and their ability to survive in diverse environments, as well as one of the ways to describe the differences between groups.

THE FOUR ESSENTIAL FUNGAL FORMS

Much of the history of mycology has used unique traits to name and sort the world's fungi—a challenging task, as many smaller fungi are hard to find and distinguish from similar looking species. To date, around 150,000 fungal species have been named,

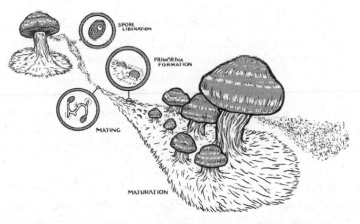

The Shiitake mushroom life cycle starts with the release of spores, which germinate mycelium after landing and branch out in search of a genetic mate. Once found, two partners fuse their tissue together, becoming one. This genetically complete mycelium then collects food and water until it can form mushrooms, which quickly mature to spread their spores.

A spore print from the Fly Agaric, **Amanita muscaria.**

a number that, though large, represents only 2–4 percent of the roughly 2.3–5 million fungal species estimated to live on Earth. (For comparison, we've named approximately 98% of the world's 350,000 plant species.)

The system traditionally used to distinguish between the many different types of mycelium-forming fungi is similar to that used for plants and animals. In general, visible features and reproduction strategies act as the major distinguishing traits between groups, with subgroups generally being based on ever-more minute details. In mycology, decades of research and debate have led to the recognition of seven fungal phyla, each of which are primarily divided by their differing reproduction strategies.

In recent years, the names and categories of fungal species have been heavily revised as improved genetics sequencing technology has shown that what were once thought to be distantly related species are actually close relatives and, conversely, that species that look quite similar actually have greatly different evolutionary histories. While this important sorting function of traditional mycology lays the foundation for applied and radical mycology—and is actually pretty fun to geek out on if you have the time to study it—it's ultimately a more technical approach to fungi than is needed to join the revolution. As an alternative to this common means of discussing the many types of mold and mushrooms, I suggest first getting familiar with the following broad categories, which don't rely on an extensive understanding of fungal biology, but still help to differentiate between the types of fungi discussed in this book.

Micro Fungi: Molds and Yeasts

The smallest fungi are the single-celled yeasts, which comprise the one percent of fungi that do not live the majority of their lives as mycelium. Yeasts are found on the surface of plants and animals and throughout the soil and oceans, where they most often act as decomposers of small pieces of organic matter. The best-studied yeasts are those that ferment liquids and leaven bread. Most other types are poorly understood, though, and the projected number of yeast species in the world continues to grow with each new study on them.

Most of the other micro fungi (i.e. species that do not form large, three-dimensional structures) show up as molds. There are likely over a million (if not two or three million!) mold species in the world, though to date mycologists have named less than 100,000. Most molds begin their life as a network of (often white, gray, yellowish, or reddish) mycelium that eventually becomes covered in a layer of colored spores. Somewhat like mushroom spores, mold spores fly off of the surface of their mycelium when a breeze or animal passes by. Unlike those of most mushrooms, many mold spores are asexual, meaning that they have the same DNA sequence as their parent mycelium. Like single-celled clones of their parent mycelium, these spores can travel hundreds of miles and, after germination, look and act exactly like the mycelial network that they came from.

Some of the most familiar molds are those that grow on old food. Many of these tend to have a not-so-pretty green or black color and funky scent. As such, most Westerners learn to dislike molds in general, and thereafter miss the opportunity to learn about the important roles that these micro fungi play in the environment and how they have influenced human history.

When given the chance, the beauty of molds can be appreciated for the subtle and delicate variations shown between species—especially in their diversity of colors and mycelial growth patterns, many of which are not expressed by their mushroom-forming kin. Molds are also incredible chemists, capable of creating some of the most important nutrients and compounds in nature, as well as a number of acids and enzymes that are used in modern food, medicine, and textile production. In parts of Asia where molds are an important aspect of fermenting traditional foods (e.g. Indonesia and Japan), the term "mold" often holds a more positive cultural connotation than is often found in the West. As with all paradigms, learning to appreciate these special traits is heavily influenced by the language we use to describe the fungi around us—whether large or small—and the lens through which we choose to view their alluring charm.

Macro Fungi: Mushrooms and Lichens
Compared to micro fungi, the mushrooms and lichens of the world often garner more respect. Like molds, the mycelium of mushrooms may grow through soil, wood, animal dung, or other organic matter. But unlike their smaller relatives, mushrooms uniquely condense their chaotic web of hyphae into a complex three-dimensional structure we call a mushroom, and which is formally referred to as a fruitbody.

Most of the soft or "fleshy" mushrooms survive for several days after arising from their mycelial network, after which time

(Left) The leaf-like thallus of **Lobaria pulmonaria,** *a common lichen species in the Pacific Northwest. (Right)* **Letharia vulpina** *is a bright chartreuse lichen that is toxic to canines and was historically used by ranchers to ward off wolves and foxes.*

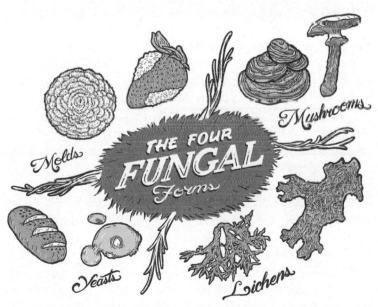

The millions of fungal species present in just four essential forms. Single celled yeasts and molds comprise the micro fungi, while the complex structures of mushrooms and the fungi-infused thalli of lichens make up the macro fungi.

they'll slowly decay if they're not eaten. Conversely, many of the firm and more wood-like mushroom fruitbodies persist for months, years, or decades, often adding an additional layer of hardened tissue to their structure during each growing season.

Alongside mushrooms, I also consider lichens to be another group of macro fungi, though this is technically inaccurate as lichens (which are symbiotic organisms and so are not categorized on the tree of life) are not entirely composed of mycelium. Rather, lichens are more like a micro-ecosystem made up of hundreds of organisms that live together in a structure primarily formed by mycelium. Roughly ninety-five percent of the total lichen structure (or thallus) is formed by the mycelial network of a single fungus, the mycobiont. Approximately 4.5% of the rest of the thallus is made of a photosynthesizing partner, or photobiont, which is usually a species of algae or cyanobacteria. The rest of the lichen is a mixed bag of hundreds of other fungi, bacteria, viruses, and microscopic animals, all living together in the "greenhouse" created by the mycobiont's mycelium. To reproduce and spread this community, some lichens form small clones, known as diaspores, which flake or fall off of the thallus. In other lichens, the mycobiont will form a mushroom that will release spores that then travel on the wind to find the other organisms that, in time, will make up another mature lichen thallus.

THE MYCELIATION OF TIME AND SPACE

Another helpful and important point of distinction between species is the role a fungus plays in the environment. The decomposing, or saprotrophic (literally, "death eater"), fungi are those that digest dead plant and animal matter, helping to kickstart the nutrient cycles that lead to healthy soils and the next generation of plants. Some decomposers eat a limited variety of substrates (e.g. some mushrooms only grow on one or two types of wood), while others have a wide-ranging appetite.

Some fungi feed off of—and may eventually kill—living plants or animals, and so are referred to as parasitic or pathogenic. Though often presented in a negative light, these fungi are important for the environment as they help limit the spread of disease, while also ensuring positive changes in the community of plants and animals in their habitat (a process known as succession).

Alongside these two groups of individualist fungi we find the collectivists that often need to live in a beneficial symbiosis with other organisms. One such group, the mycorrhizal fungi, grows on the surface and inside of plant roots, where they exchange water, nutrients, and minerals gathered from the soil for sugars produced by the plant during photosynthesis. Another

Fungi shape the world around us on geological time scales vastly greater than the short lifespans we use to frame human history. Silent stewards of Earth, they guide and contain all that we hold dear.

group, the endophytic fungi, symbiotically live inside of the tissue of all plants (leaves, branches, and all!), where they support their partner's health and protect the plant from environmental stressors, such as heat or drought.

Many fungi blur the lines between these categories and might perform two or more ecological roles simultaneously, or change their job at different stages in their life. For example, a root-associating fungus might also decompose organic matter, while a species that lives inside a plant's leaf may become parasitic inside the body of an insect that eats that leaf.

In a healthy ecosystem, the countless fungi present work in concert to perform specialized roles that complement each other, while also ensuring that the health of their habitat is sustained for centuries. Just as an individual human can profoundly influence the health and growth of their larger culture, so does every hypha add its own unique insights to its greater network, while also dynamically engaging with the world it depends on for survival and self-discovery. The implications of these and other aspects of fungal ecology are vast, as together they strongly suggest that these so-often disregarded species have been critical for the sustainment of all life on Earth since the earliest epochs—a conclusion that begs the question of what else has been overlooked and what else have we forgotten?

TERMS TO KNOW

Like all sciences, mycology carries its own unique set of terms—so many, in fact, that whole dictionaries are dedicated to them! Below are some of the most helpful and commonly used myco-words I suggest becoming familiar with. More words are also defined in the Glossary on page 187.

- **Fungi** (*sing.* **Fungus**): Non-photosynthesizing eukaryotes that digest their food externally. Most produce, and live inside, a network of branched tubes (hyphae) that grow from their tip, while a small number live as individual cells.

- **Hypha** (*pl.* **Hyphae**): The primary, tube-shaped tissue of almost all fungi.

- **Mushroom**: The macroscopic spore-bearing fruitbody of a fungus, often (but not always) hosting a stalk and cap in various sizes, colors, and degrees of ornamentation.

- **Myceliation**: The growth of mycelium over and through a substrate.

- **Mycelium** (*pl.* **Mycelia**): A collective network of a fungus's hyphae; the vegetative structure of a fungus.

- **Spore**: A specialized fungal reproductive structure, often single-celled and designed to be dispersed and travel away from its parent mycelium.

- **Substrate**: 1) The food of a fungus. 2) A substance acted on by fungal digestion. 3) The material from which a fungus produces a fruitbody.

- **Yeast**: Single-celled fungi that live in a variety of environments, often acting as decomposers.

The Past is Myceliated

To sense where we are going with fungi—or, perhaps, where they are leading us—we must first understand where the mycoculture has come from. This is true for humanity on the whole, as well as for each individual engaged in the current revolution. Learning how our ancestors—whether from an inherited or chosen lineage—interacted with the fungi of their environment provides a sound starting place for orienting a compass down the mycelial path.

Embracing traditional skills that honor wild fungi takes many forms. If you are able to trace back and identify with your family's ancestry, learning about or practicing the myths, customs, and cuisines of that lineage is one starting place for developing a connection to the mushrooms that you identify with. Across the inhabited continents, traditional cultures found numerous ways to revere fungi, whether through rituals devoted to the fermentation of alcoholic drinks, in the consumption of mushrooms (a practice dating back at least 18,700 years), or by honoring the abundance of the world's ecosystems, which fungi fundamentally sustain. Where these customs are still known, they can be brought into holidays and celebrations throughout the year, whether as originally practiced or in an updated and recontextualized form.

Learning about the diversity of mushrooms and microfungi in your most cherished land base—as well as the traditional ways to engage with them—is another route to spawning a connection to mycology. Fungi are all around us, no matter the season. They permeate plants and soils, just as their spores fill the air and are taken in with our breath. The simple act of recognizing the impacts of mushrooms and molds in our lives and on the land—an experience I refer to as seeing fungi—offers its own approach to shifting our perceptions of the day's challenges, simply by attending to the often-overlooked beauty of some of the natural world's more subtle aspects. In time, this practice lends to a sense of what it means to live like fungi: to slow down, provide support to others, and uncover answers to our greatest problems through the efforts of a patient inner resolve.

Looking to our own bodies, we find a means to connect with fungi in learning about and supporting the beneficial species that live inside of us and influence our health. This community, known as the mycobiome, complements the microbiome of bacteria and other organisms that enhance our digestion and, as research is finding, influence our immune response and mental state. By following the latest discoveries in this important field, each of us may, in the coming years, be able to address currently unresolvable health issues through the simple act of taking *pro-mycotic* supplements (i.e. inoculum of helpful fungi that our bodies may lack) or *pre-mycotic* substances that feed the friendly fungi inside of us. If fungi prove to be an essential yet entirely overlooked aspect of human health, this most intimate form of human-fungal relations will likely also be one of the most impactful on social acceptance of the innate value of these powerful organisms.

Developing a personal practice for engaging with the healing benefits of mushrooms and other fungi—whether as nutrient-dense foods, natural medicines, or silent guides in the forest—is among the most potent forms of self-care I have encountered. The more I work with fungi and feel their influence throughout my life, the more I am drawn to imagine what an ancestral reverence for the natural world may have felt like, as well as how I might best reflect those practices through a rootedness with place that modern technology all but erases from view. As they remind me to cultivate a sense of stillness, they also infuse a resilience within me to face the challenges of today and the uncertainty of tomorrow. As I contemplate their many mysteries, I am inspired to seek new discoveries in the unknowable worlds around and within myself.

Current models of evolutionary theory suggest fungi or similar organisms were the first complex-celled species on Earth, meaning that plants and animals descended from them. Fungi are our greatest ancestors, whose wealth of wisdom has been a constant force throughout time—and that permeates and shapes all environments today. As research increasingly proves that much of what we revere about the natural world stems from the unique actions of fungi, the greater the space will be in the mycoculture to dance with wild mycelium like nobody's watching, while calling out to everyone to join you on the forest floor.

We are fungi. We are their symbiotic partners, their descendants, and their dependents. Deeply seated in the human condition, their impacts on history are widespread and profound, yet unfortunately unknown by most. As the modern mycoculture

rises and new connections are built between humans and hyphae, recalling and redefining this legacy is the first and most essential step toward embodying the many gifts of a much-needed revolution. As we each play our part in this process, we're given the opportunity to shape the future in an image of our own making. How this story continues to unfold is not to be determined by any one person, but by all involved, as informed by the unique lessons they've learned and stories they crafted by working with the fungi of their lives.

Chapter 1: FORAGING FOREVER

$\mathcal{S}$ince the earliest civilizations, the most direct path to forming a personal relationship with fungi has been through the gathering of wild mushrooms. Whether to source ingredients for a meal, medicine, or ceremony, early humans spent generations testing the edibility and applications of the mushrooms near them, eventually determining which species were to be cherished, and which were best left where found. Over the millennia, traditional healers and herb wives refined this knowledge until it was first documented in Western texts in the 1700s. Since that time, the naming of fungi has become increasingly nuanced as more species have been discovered and described, and yet the skills needed to find the mushrooms nearest you have changed little since the first fruitbody was picked by humans so long ago.

Today, around 14,000 mushroom species have been cataloged globally, with likely thousands left to be named in the coming years. Thankfully, learning to identify all of these species is not needed to safely forage mushrooms. By starting with a small number of the easier-to-recognize species (or "foolproof fungi," FPF) in your area, you can begin foraging with little practice. Many people only learn a handful of their local FPF and find great joy in that slimmed down approach, while a smaller percentage of foragers go on to learn the names and detailed taxonomy of hundreds of species in their region.

While many people enjoy foraging as their first step into the world of mycology, others, such as myself, come to the practice from the worlds of mushroom cultivation or medicinal mushrooms. As with other aspects of mycology, no matter how a person enters the science, I find that in time, one way of working with fungi will lead to a curiosity about all of their roles in the world.

Regardless of how you come to the art of foraging, the edibility of your local species is likely to be an initial consideration. Wild edible mushrooms are not only a free and delicious food that can be dried and savored year-round, but also a nutrient-dense protein and mineral source that often provides potent healing properties. In addition, some of the most prized gourmet mushrooms cannot be cultivated in farms and must be wild harvested. This rarity not only offers their discovery in the wild to be cause for celebration but, for some, a means for substantial income as the wild mushroom market is one of the largest (legal) cash-based industries in the world, with an estimated total value of over one billion U.S. dollars annually.

Along with these self-sufficiency-increasing benefits, I also find great joy in the unique means that mushroom foraging provides for connecting with my bioregion and for tuning into the patterns, climate, and ecology of the environments where fungi are found. By slowing down to scan the forests and fields, foraging enables us to see the spaces we walk through with a personalized lens—one that is guided by the rhythms of the wild and shaped by the nooks of each hillside and valley. In time, learning to see fungi in all of their forms becomes second nature and ultimately a practice that is carried wherever you travel, with each new species encountered adding its own special means for appreciating the beauty of fungal diversity and the mysteries of the natural world.

FIND YOUR MYCOFOLK AND THE MUSHROOMS WILL FOLLOW

If you are new to foraging, the easiest way to get oriented in the practice is by joining the mycological society, or "mushroom club," nearest you. These groups are primarily focused on teaching others how to identify popular edible and toxic mushroom species, as well as when and where those species are most likely to occur in your part of the world. In the U.S. and Canada, the North American Mycological Association (namyco.org) oversees many of the continent's formal clubs, and most continents and countries have similar organizing bodies that can be found with a quick web search. Complementing these formal groups are a growing number of regional mushroom hunting groups on Facebook that help locals identify mushrooms through photos.

Joining a club not only builds confidence in your identification skills by learning directly from helpful mentors, but also provides unique insights into the fruiting patterns and ecological influences of the mushroom species nearest you. This place-based knowledge is rarely recorded, and so is in constant need of careful stewarding between generations of mushroom seekers. Being surrounded by other mushroom enthusiasts is also encouraging, especially if you don't currently know other people who share your interest in fungi. Likewise, contributing to the local mycoculture of your mushroom club might lead to the development or rediscovery of local customs and cuisines related to the most culturally important mushrooms in your bioregion.

If you don't have access to a nearby club, taking an in-person or online foraging workshop from a reputable teacher or getting a well-written and regionally appropriate mushroom field guide are alternative means to start learning. Ideal teachers do not need to hold degrees in mycology, though they should at least have multiple years of experience foraging and identifying wild mushrooms in your part of the world and demonstrate a clear concern for your safety alongside demonstrable knowledge of identification techniques.

Be sure to also seek out the most up-to-date field guide for your area, which would ideally have an organizational system that makes it easy to use, while also providing multiple color photographs of each mushroom described. If possible, acquiring guides from different authors is also helpful, so that you have several depictions and photographs to compare.

To help you make the most of your field guide, the rest of this chapter is written to lower your learning curve as quickly as possible, while complementing your field guides, which are indispensable for ensuring accurate identifications.

ON PLANNING, SAFETY, AND ETHICS

The art of mushroom foraging can be as carefree or as methodical as you like. Many people like the simplicity and excitement of going to new locations during the mushroom season and carrying nothing but a knife and a basket as they hope for a bumper crop of one of their favorite edibles. Other foragers spend several seasons building up a map and calendar of the most productive foraging sites, so that future outings are shorter and more certain to bring back a large collection.

In between these two approaches is the best-guess method, where, after choosing the species you would like to find, you refer to local field guides to determine when and where that mushroom is most likely to occur near you. For example, if your local Chanterelles (a popular edible species) typically occur from August–October in coniferous forests (like they do in the Pacific Northwest), you'd know not to look for that species in a predominantly hardwood forest in the spring.

However you decide to plan (or not plan) your trip, the first step before leaving is to ensure your safety in the field. Along with the standard risks that can come with going on a hike (e.g. a twisted ankle or an unexpected downpour), foraging for mushrooms adds the heightened chance of getting lost in the woods due to the hours spent looking at the ground and wandering aimlessly off trail. Telling others of your travel plans and carrying a compass, charged phone, and proper outerwear are easy steps to ensuring you'll make it home safely after each foray.

To avoid consuming a poisonous mushroom, it is best to place each collection (i.e. each group of mushrooms harvested from the same immediate area, and which you are fairly certain are the same species) into a separate waxed or paper bag (plastic bags make mushrooms sweat and thus go bad quite quickly) and to only consume a mushroom if you are *absolutely certain* of its identity. This last point can't be emphasized enough as there are many species that can make you very sick and a small number that are deadly (though there are more deadly poisonous plant species in the world than deadly poisonous mushroom species). Luckily, the risk of being poisoned by mushrooms can be greatly avoided by only consuming species that are easy to recognize and that do not have poisonous look-alikes (i.e. the FPF near you!). Likewise, double or triple checking the identity of a mushroom by comparing descriptions between multiple field guides is well worth the time it takes as it further ensures you do not mistake the identity of a poisonous species for an edible one.

A final consideration is the impact your harvesting might have on the environment. While research suggests that the act of picking mushrooms on its own doesn't discourage future fruitings (just as picking an apple doesn't reduce its parent tree's yield the following year), disturbing a mushroom's habitat by compacting the soil or trampling nearby plants can quickly turn a bountiful patch into a barren plot. If you are in a small group and go off trail, you can avoid trampling by walking in a single file line. If you're in a larger group, spread out. Lastly, help ensure the peaceful enjoyment of the next forager's hike by packing out any items you bring into the wood, covering the stumps of the mushrooms you picked, and moving any discarded mushrooms far from a trail's edge.

On Collecting Lichens

Foraging lichens is somewhat similar to picking mushrooms, with the need for a good field guide and a sense of where and when to look being helpful tools for planning. The greatest difference between a mushroom and lichen foray, though, is that removing lichens from an environment should only be done when there is a justifiable reason for doing so. Unlike mushrooms, harvesting a lichen removes the entire organism and leaves almost nothing behind to propagate that species. Likewise, most lichens grow very slowly—with some only growing one millimeter per year—so removing a large specimen might represent the end of a decades-long history of that being.

The rule I have given myself is that I will only collect a lichen if 1) there is an overall abundance of that species in a given habitat, 2) I am certain that it is not a threatened or endangered species, and 3) I have a clear intention for working with that lichen, such as to relocate the collection from an area destined to be deforested to a habitat where it will be able to continue living.

TOOLS FOR THE BODY, BASKET, AND BACKPACK

Listed below are some of the most helpful items to take on a foray. Many of the smaller tools can fit into a waist pouch, while a small hiking backpack will hold all of these items if you want to be fully prepared.

Use

> E - Essential
> H - Helpful
> S - Safety

On Your Body

- Bright clothes and sturdy footwear – E
- Compass – E / S
- Whistle – E / S
- Headlamp – H / S
- 10–30x hand lens (jeweler's loupe) – H
- Charged phone with GPS capabilities and camera – H / S
- Gaiters – H
- Orange vest or hat – S
- Bear spray – S

In Your Basket

- Weather-appropriate note-taking materials – E
- Knife and small paintbrush – E
- Paper or waxed paper bags of various sizes – E
- Tackle box (for sorting small mushrooms) – H
- Trowel – H
- Vegetable peeler (for cleaning off dirt) – H

In Your Backpack

- Binoculars – H
- Chisel (for harvesting woody mushrooms from logs or snags) – H
- Emergency blanket, bivvy, or poncho – S
- Identification field guides – H
- First aid kit – S
- Food and water – S
- Insect repellant – S
- Map – H / S
- Sunscreen – S
- Tick removal tool – S
- Waterproof rain gear – S

I generally keep a small pouch of the essential foraging tools with me no matter where I am, so that I am ready to collect a mushroom while on a bike ride or at a neighborhood park. If I am going on a longer foray, I will fill a thirty-liter hiking backpack with all of the above items, as well as specialty tools for collecting mushroom tissue (for cultivation purposes) and spore prints, and for taking close-up photos of the mushrooms in the wild. While these tools help add more layers of engagement with the mushrooms I work with, the most important aspect of the entire process is ensuring that I am present and engaged in the habitats I travel through, all in hopes of meeting new fungi friends, or of seeing the return of old acquaintances.

FOOL-PROOF FUNGI (FPF) AND DEADIBLES

As with plant foraging, the first mushrooms to learn are the deadly poisonous species in your area (the "deadibles"). These vary somewhat by region, though all North American foragers should be aware of *Galerina marginata* (a small, brown mushroom that looks like many other brown mushroom species), and the deadly *Amanita* species (i.e. the Death Cap [*Amanita phalloides*] and

Destroying Angel group [*Amanita virosa* and its close relatives]), as these are among the most commonly consumed lethal mushroom species. Your local field guide should note if these or similar poisonous species are nearby, as well as how to identify them. In general, any small to medium-sized mushrooms with primarily brown features should be avoided by beginners, as should all *Amanita* species, which grow out of the ground, produce white spores, have free gills, and have a cup or bulbous feature known as a volva at the base of their stalk (though this may be buried underground).

After your local poisonous species are determined by reading your field guide, the next mushrooms to look for are the edible FPF of your area. As with poisonous species, FPF vary by

Galerina marginata *and a number of* Amanita *species are most commonly connected to lethal mushroom poisonings. Beginning mushroom foragers should not consume any mushroom that looks similar to these species. Refer to your regional field guide for details and photographs of local poisonous varieties.*

region, though below are seven of the most widely distributed of these mushrooms, all which are described in greater detail at the back of this book.

- **Chanterelles** (*Cantharellus spp.*)
- **Chicken of the Woods** (*Laetiporus spp.*)
- **Morels** (*Morchella spp.*)
- **Shaggy Mane** (*Coprinus comatus*)
- **Giant Puffball** (*Calvatia gigantea*)
- **Cauliflower** (*Sparassis spp.*)
- **Maitake** (*Grifola frondosa*)

(GREAT) NOTES AND PHOTOS WILL SAVE US ALL

If you think you've found one of the above FPF, take a minute before picking it to write down its color, scent, size, habitat, and what it's growing from in your notebook, then take several photos of the mushroom in its surroundings along with a few close-ups of its features. The more information you collect in the field, the more you will have to refer to later on when reading your field guide's species descriptions.

If you choose to pick the mushroom, be sure to harvest its entire fruitbody, including any underground pieces of the stalk,

Poisonous **Amanita** *species start as an underground "egg" (known as a universal veil), out of which the mushroom emerges.*

and then place it into its own paper bag. Write a unique collection number on the bag as well as next to your notes, so that it will be easy to correlate everything when you return home.

Using your notes and photos, navigate through your field guide's identification process (which will vary by guidebook and should be explained in your guide's opening pages). If you think you have one of your area's FPFs, compare one guide's description of that species with your notes and *all* of the features of the mushroom you collected, as well as with the description of another field guide (or two!). Once you are certain of the mushroom's identity, enjoy your harvest and the knowledge that mushroom foraging and identification isn't that hard if you give it a little patience!

CURATING COLLECTIONS

After you've identified your collections as best as you can, the next step is to either discard, preserve, or process the

The author with a Cauliflower mushroom found along a trailside at a public park in Portland, Oregon

mushroom. If you don't want to keep the mushroom, it's ideal to return it back to where it was harvested from, so that it can feed back into that habitat's nutrient cycles. If that's not possible, composting the mushroom is a good second option, so long as the mushroom isn't poisonous (which avoids any chance of toxins making their way into someone's dinner plate). A third option is discarding the mushrooms in a nearby park or other natural space, so long as they are out of sight of future visitors.

If you wish to preserve a mushroom as a reminder of your foray, drying it should be all that is needed. Many individual collectors maintain a *fungarium* (the mycological version of an herbarium) in their homes—whether in temperature-controlled cabinets, or on display across mantles, altars, and bookshelves.

The best practice for preserving most mushrooms is to dry the fruitbody at no higher than 130°F (55°C) until it is crisp like a cracker, and then place it in a freezer for two nights to kill off any internal insects that might otherwise eat the dried mushroom. It is best to dry your collections first as freezing a fresh mushroom

*A day's bounty of Lobster mushrooms (**Hypomyces lactifluorum**) and Chanterelles (**Cantharellus cibarius**).*

can cause it to lose structure after thawing due to its cells swelling and rupturing while freezing. The mushroom can then be placed in a storage container along with its identifying information and helpful notes about where and when it was collected.

Some of the more important mushrooms I've come across are stored in a fungarium at MYCOLOGOS, while many others decorate my living spaces to remind me of the joys of special forays gone past. Even during the humid winters of Oregon, most of these mushrooms survive unprotected with no signs of degradation. Occasionally, a woody species will attract small insects that burrow through the fruitbody, creating a small pile of dust below the mushroom. When this happens, I put the mushroom in the freezer for a few days to fend off the insects.

The third route to working with your collection is to process it into food, medicine, paper, or dye for hair or fabrics, or to cultivate it in or around the home—techniques that are covered in the coming chapters!

Safe and efficient mushroom foraging starts with picking a good time and place to go and packing all of the necessary gear. In the field, each find should be thoroughly documented, so that the identification process is as accurate as possible later on. Processing the collection may include dining, dehydrating, or discarding, depending on the species.

Chapter 2: UMAMI, THAT TASTES GOOD!

*A*longside foraging for mushrooms, making fungi-infused dishes and drinks are among the most traditional and culturally significant means for working with edible species. Mushrooms are found throughout many traditional Asian, African, European, and North American dishes, just as the fermenting actions of many molds and yeasts underscore the customs and cuisines of various cultures.

As a food source, fungi offer a wide range of flavors, with many mushroom species tasting fruity, nutty, earthy, or even sour. Most of the more popular edible mushrooms, such as Shiitake, Maitake, and Morels, are revered for their unique, complex, rich, dense, and savory taste that, while often compared to meat, are taste sensations that are truly only found in fungi.

These flavors stand out on our tongue due to the presence of compounds that stimulate *umami* receptors in our mouth. This "fifth flavor" complements the more familiar salty, sweet, bitter, and sour tastes and is attributed to the filling and nourishing sensation produced by fatty or protein-rich foods, such as animal products, and a small number of vegetables, including seaweed, tomatoes, and eggplant. Umami has such a strong effect on our perception of food that the umami receptor stimulating compounds like monosodium glutamate (MSG) are added to many processed foods to trick our brains into thinking that a dish is more filling than it is.

Along with mushrooms, several mold species also produce strong, umami-rich flavors—as experienced in fungally fermented foods like tempeh, blue cheese, and bonito fish flakes. The umami of micro fungi was discovered by various traditional peoples millennia ago, and remains cherished generations later due to the unique, transformative, and complex flavors that only these fungi offer. Today, myco-literate chefs are experimenting with traditional fermenting fungi—such as the koji mold (*Aspergillus oryzae*) that has long been used to make miso and tamari—to grow tasty, moldy versions of common dishes with a modern twist. As this trend continues, it's likely that new fungi-infused ferments will make their way to grocery store shelves in the not-too-distant future to provide more flavorful versions of some of today's most mundane ingredients.

Complementing the rich flavors of mushrooms are the variety of soft, gelatinous, tough, and chewy textures that their fruitbodies offer. Additionally, most gourmet mushrooms (e.g. Shiitake, Oysters, Chanterelles, and Truffles, to name a few) are nutrient-dense, with many cultivated species being high in protein and trace minerals, as well as vitamins A, C, or K. More uniquely, the ergosterol in the cell walls of fungi (discussed in the Introduction) converts to vitamin D_2 upon exposure to strong ultraviolet light. This means that you can place any fresh or dried mushroom in the sun for a few hours and increase its vitamin D content! Fungal ferments also transform the nutritional value of the ingredients that they are made from, often increasing the concentration of various vitamins, while also improving the digestibility of the food overall.

For all of these reasons, mushrooms and other fungi have been a staple in many traditional diets for thousands of years. Today, Mexico, Guatemala, China, and Japan are among the world's most fungi-hungry countries, with hundreds of mushroom and micro fungal species playing important roles in their local cuisines.

Despite all of these benefits, many people in the West today think of mushrooms as having an unfavorable flavor or texture—a phenomena I attribute to a general lack of well-prepared mushroom dishes in most Western restaurants and to the prevalence of Portobello and Button mushrooms, which make up the largest percentage of the cultivated mushroom market.

These mushrooms are gastronomically underwhelming relative to species with more complex flavors, and, in my experience, are often under-cooked by both professional and amateur chefs—such that the mushrooms end up being served overly wet and with a slippery texture. Is it no wonder, then, that most people in the West don't like to eat mushrooms?

ELEVATED TASTES FOR FRUITING BODIES

To address this unfortunate state of fungal affairs, below are a few of my favorite dishes that not only highlight the diversity of ways to work with fungi as food, but that are also fun to make!

> Key:
> Gluten-free – GF
> Keto-friendly – K
> Paleo-friendly – P
> Vegan-friendly – V

The Simple Sauté – GF / K / P / V

My go to method for preparing an edible mushroom, especially if I have never eaten it before, is to create a simple side dish that brings out that species's unique flavor and texture. First, I coarsely chop the mushrooms and then dry sauté them over medium heat (this means adding them straight to a hot pan without any oil at first). The water in the mushrooms will start to boil out, creating steam, while shrinking the mushrooms down. This concentrates the mushrooms's flavor and also creates a firmer texture. Once the steam stops coming out of the mushrooms, I add enough oil to fry the fruitbodies until they are dense and almost crisp, and then add salt to taste and, occasionally, a pinch of pepper. This simple process not only enhances a mushroom's flavor, but also reduces the soft-and-slippery texture that many dislike. It's so effective that I have been able to convert several people that swore they did not like to eat mushrooms into raving mush-aholics!

When dry sautéing, using a spatula to push out the water within the fruitbodies helps shorten cooking times.

An excellent first step to enjoying edible mushrooms is to steam off their internal water over low heat before adding oil and frying them to your preferred consistency. Then just add some salt and enjoy!

Safe MyConsumption

Apart from the mushroom species that cause illness in almost anyone that eats them, there are also a number of wild harvested species that are safe to eat by most people, and yet cause stomach upset in a small percentage of the population. In addition, some people can have negative reactions to some of the most commonly consumed mushrooms. For these reasons, it's best to only eat a small (approximately one tablespoon) amount of a cooked mushroom the first time you try it and to then wait a day to see how your body reacts before eating more of that species. It is also a good idea to not eat a large amount of any mushroom species in one sitting or a mix of many species at once, as this tends to give many people some degree of stomach upset. Lastly, *all* mushrooms should be cooked prior to eating as heat not only increases the digestibility of mushrooms, but also destroys any harmful compounds, microbes, or insects that may be present in the mushrooms.

King Oyster Scallops – GF / V

The King Oyster mushroom (*Pleurotus eryngii*) is a dense-textured species with a long, thick stalk. When sliced into rounds and fried in a marinade, this mushroom readily takes on the appearance and texture of sea scallops, albeit with a notably fungal flavor!

Ingredients

- 1 lb. or more King Oyster mushrooms
- 0.5 C. Water
- 0.25 C. Tamari
- 3 Thin slices of fresh ginger
- 6 Garlic cloves, halved
- 2 6-inch Pieces of kombu seaweed

- 2 T. Miso paste
- High-temp oil as needed

Recipe

1. Trim dirt off of each mushroom's stalk base if needed, then slice the entire mushroom crosswise into a series of one-inch thick discs.

2. Mix the water, tamari, ginger, garlic, kombu, and miso in a saucepan, and simmer the mix for 20 minutes, creating a marinade.

3. Score the mushroom discs with several crosshatch cuts, then set them in an even layer in an oiled pan over medium-high heat.

4. Drizzle the miso marinade over the discs as they sear for several minutes, adding more marinade as is needed to ensure that the mushrooms retain much of the sauce's flavor.

King Oyster mushrooms fruiting at MYCOLOGOS

5. Once browned, flip the discs over and repeat for the other side.

6. Serve immediately with a side of seared vegetables or over your favorite style of pasta.

Nameko Nabemono - P / V

Nameko (*Pholiota nameko*) is a nutty-flavored mushroom that is best known for the unique, sticky texture of its cap's surface. A versatile mushroom, Nameko is well suited for soups, sauces, stews, risottos, pizzas, and pastas, while also pairing well with poultry, steak, fish, tofu, nori, cabbage, olives, quinoa, miso, sake, goat cheese, daikon radish, and pinot noir. Nameko is

King Oyster stalk rounds sliced and ready for cooking.

King Oyster scallions ready with a side of grilled asparagus.

regarded as one of the more nutritious gourmet mushrooms, as it hosts over a dozen amino acids, as well a good amount of potassium, magnesium, phosphorus, iron, zinc, copper, selenium, molybdenum, and the vitamins D, B1, B2, B3, B6, B7, and B9. Medicinally, Nameko has shown liver supporting, immune boosting, anti-inflammatory, and antitumor effects, and it may even support people with high cholesterol.

In Japan, Nameko is traditionally served in sushi rolls or in a mixed pot of vegetables, meat, and broth known as *nabemono*. A basic version of this dish is described below, but feel free to adjust it to taste.

Ingredients

- Nameko mushrooms, chopped
- 1 packet of Soba noodles
- 1 packet of Split-bean natto
- 2–3 T. Daikon radish, grated
- 1–2 T. Green onion, chopped
- 0.5 C. Mentsuyu
- 0.5 C. Water

Recipe

1. Cook the soba noodles and drain. Set them aside for now.

2. Grate the radish and place it in a colander to drain off any excess moisture.

3. Mix the natto until it is very sticky.

4. Mix water and mentsuyu to taste and bring the mixture to a boil in a small pot. Add the Nameko mushrooms (they can first be sautéed using the method from the Simple Sauté recipe if desired) and reduce the pot's heat to a simmer for 5–10 minutes. Add green onions and turn off the heat.

5. Place the noodles in a bowl and cover them with the natto and grated daikon.

6. Pour the Nameko sauce over the bowl.

7. Enjoy!

Lion's Mane Latkes – GF / P

Lion's Mane (*Hericium erinaceus*) grows as a mass of thin, white "teeth," making its fruitbody look similar to a pom-pom. Despite its unusual appearance, this mushroom is revered for its excellent crab-like flavor and texture, and powerful benefits for the nervous system. In addition, it is made up of more than 30% protein, while also offering a good amount of phosphorus, iron, calcium, potassium, magnesium, thiamin, riboflavin, calciferol, and niacin. Thankfully, the mushroom has become increasingly cultivated in the West and may very well be available at a nearby farmer's market if you aren't able to find it at your favorite grocery store or

A Lion's Mane fruitbody, ready for the skillet.

food co-op. It's an excellent, go-to mushroom in many dishes and one well worth trying.

Ingredients

- Lion's Mane mushrooms, diced
- 2 Large potatoes, grated
- 1 Large onion, grated
- 2 Large eggs
- 0.5 C. All-purpose gluten-free (or paleo) flour
- 1 t. Baking powder
- Butter or high-temp oil
- 2 t. Salt
- 0.5 t. Pepper

Recipe

1. Combine the grated potatoes and onions inside a colander, then wring out their excess moisture before setting the mix aside to drain.

2. Dry sauté the Lion's Mane pieces until steam ceases to be released, then set them aside to cool.

3. Combine the drained potatoes and onions with the mushrooms, eggs, flour, baking powder, salt, and pepper.

4. Using a spoon, place small balls of

batter in the pan. Flatten the balls with the spoon, then fry and flip the discs until they're crisp on both sides.

5. Set the finished latkes between two paper towels to cool off and to remove excess oil.

6. Season with salt and pepper to taste and serve while warm.

Love Mushroom Tempura – V

The Love Mushroom (or Pink Oyster, *Pleurotus djamor*) is a small, clustering mushroom with a bright pink color. Some people find its flavor not as rich as the related Pearl Oyster (*Pleurotus ostreatus*), making the use of tempura's fried batter coat an easy way to add a bit more nuance to the mushroom, while also enabling their pink color to surprise guests after they take a bite.

Ingredients
- 12–20 Love Mushroom caps
- 1 C. Tempura flour
- 0.75 C. Cold water
- Green chives
- Salt and pepper to taste
- Frying oil
- Tamari
- Pickled ginger

Recipe
1. Mix the flour, water, chives, salt, and pepper together with a whisk or fork while you warm a quarter-inch of oil in a pan over medium heat.

2. Dip the mushroom caps in the batter (you can add them fresh or dry sauté them first, following the Simple Sauté recipe above), then fry them in the oil until they're crisp and brown.

3. Set the finished caps between two paper towels to cool and to remove excess oil.

4. Serve while warm with a dipping bowl of tamari and a side of pickled ginger.

Candy Cap Cookies – V
Candy Cap mushrooms (*Lactarius camphoratus*, *L. fragilis*, and *L. rubidus*) are among the few species best used in desserts as they produce a rich scent and flavor when dried that is akin to maple syrup, butterscotch, and caramel. If these mushrooms don't grow near you, they are likely available to order online—an effort that is well worth making. Candy Caps can be made into ice cream, infused in whipped cream, or added to the secondary fermentation stage of beer making. A popular approach is to also mix them into cookie batter. If you have a favorite cookie or brownie recipe, you can simply incorporate pieces of the dried mushroom to make the baked goods decidedly fungal. Below is a simple cookie recipe that I like to use whenever Candy Caps cross my path.

Ingredients

- 1 C. Dried Candy Cap mushrooms
- 1 C. Butter or oil (to make it vegan), softened
- 1 C. Sugar
- 1 Egg
- 0.5 t. Vanilla
- 0.25 t. Salt
- 2.5 C. Your favorite all purpose baking flour, sifted
- 0.5 C. Toasted pecans, chopped

Recipe

1. Place the dried mushrooms in just enough warm water to cover them and allow them to rehydrate until soft (approximately 20 minutes).

2. Preheat your oven to 350°F (175°C).

3. Gently squeeze out excess water from the mushrooms and set the favor-filled water aside for later.

4. Chop and lightly sauté the mushrooms (as described in the Simple Sauté recipe) until your preferred texture is achieved.

5. Mix the butter and sugar until a cream-like consistency is reached, then mix in the egg and vanilla.

6. While stirring, slowly add the flour, chopped nuts, and cooked Candy Caps.

7. Roll the dough 0.25-inches thick, then use your favorite cookie cutter to cut the dough up.

8. Place the cookies on an ungreased cookie sheet and bake them for 8–10 minutes until their undersides are golden brown.

FUNGAL FERMENTS

Compared to cooking mushrooms, the edible micro fungi each require a bit more attention to bring out their fullest flavors. Thankfully, though, after generations of working with these fungi, the best practices have been refined to the point that most fungal ferments can be made with not much more than a few common kitchen tools and a good degree of patience.

Soy-Free Tempeh – V
Tempeh is a traditional Indonesian "cake" of soybeans that have been fermented and bound together by the mycelium of *Rhizopus oligosporus*. A moldy bean patty by another name, this protein-rich staple hosts a nutty flavor, crunchy texture, enhanced digestibility, and increased nutritional value when compared to plain soybeans, all of which is brought about by the transformative actions of the fermenting hyphae.

Tempeh is traditionally made by rubbing banana leaves covered in the mold on partially cooked soybeans, which are then wrapped in a leaf to ferment for a day or two as the fungus rapidly grows through the beans. In North America, where the mold is not found in the wild, starter packets of the fungus's spores can be bought online and sprinkled on soybeans or other legumes or grains to create soy-free versions of this meat alternative. Below is a general recipe for making this modern fungal ferment, which can be adjusted to match your ingredients preferences.

Ingredients
- 1 Packet of tempeh starter (usually made of spores diluted in a grain or bean powder)

- Substrate of your choice. Popular options include black beans, chickpeas, hemp seeds, lentils, mung beans, quinoa, and lima beans.

- Rice vinegar, white vinegar, or pasteurized apple cider vinegar

Equipment
- Clean incubator, dehydrator, warm closet, cupboard, or attic space
- Cutting board
- Fork or Knife
- Large mixing bowl
- Large spoon
- Plastic sandwich storage bags

Recipe
1. Cook the beans for your base until they are just under-cooked for normal consumption. If they are very firm, the final tempeh will not be enjoyable to chew, and if the ingredients are over-cooked the mycelium won't be able to bind the material very well. This is the trickiest part of the whole process, so consider doing a few small test batches to dial in your substrate's ideal cooking time.

2. Strain the substrate and allow it to cool to around 75–80°F (24–27°C).

3. Clean your table, cutting board, and utensils thoroughly and set them on a clean surface until you need them.

4. Read the instructions that come with your tempeh starter to determine the ratio of starter to cooked substrate.

5. Using a clean spoon, sprinkle the starter across the substrate and stir to mix them evenly.

6. Lay several sandwich bags on the cutting board and use your knife or fork to perforate the bags, creating a grid of small air holes spaced evenly every one inch (2.5 cm) or so.

D.I.Y. TEMPEH

STARTER

BEANS OR GRAINS

VINEGAR

PERF. BAG

24 HRS @ 85-91°F

① MIX & PACK   ② INCUBATE   ③ SLICE & COOK!

This mold-based dish is made by sprinkling spores on partially cooked beans, and then incubating the mix in a perforated plastic bag for 24–48 hours at 85–91°F.

7. Spoon the inoculated substrate into the perforated bags until a 0.5–0.75-inch (0.2–0.3 cm) layer is created, then seal the bag. Lay the bag flat and tamp the mix to create an even patty.

8. Place the filled bag in a warm space that can be maintained at 85–91°F (30–33°C), so as to germinate the warmth-loving spores. A countertop dehydrator that does not get too hot, or a home built incubator made from an old mini fridge are popular DIY incubator options.

9. Periodically check the bags over the next two days. If all goes well, the *Rhizopus* mycelium will have covered the substrate, forming a fluffy white cake that smells sweet and nutty. If the mold grows for too long, it will start to produce black spores, which are safe to eat, but not very visually appetizing. If the cake becomes green or takes on a scent that is at all nauseating or unappealing, it has likely been contaminated. If this occurs, simply start the process over, this time paying extra attention to ensuring that your kitchen and tools are as clean as possible.

Mushroom Mead – GF / V

One of history's most popular fungal ferments are the alcoholic beverages produced by yeasts (such as *Saccharomyces cerevisiae*, which creates most alcoholic drinks today). After centuries of development, the multi-step process of making high quality beer and wine (e.g. roasting barley, simmering hops, or mashing fruit) is often intimidating for beginners wanting to make their first brew. Thankfully, making a traditional mead is an easy starting place for the novice as it only requires mixing honey and water in a clean container. This classic drink can be modified in many ways, such as by including fruit or spices to the mix, and the amount of honey used can be adjusted widely to get varying alcoholic concentrations and degrees of sweetness. Below is a simple mead recipe that incorporates mushrooms as a flavor addition—each step of which offers wiggle room for experimenting if you wish.

Ingredients

- High quality honey that is ideally raw (unpasteurized) and locally produced
- Champagne or mead yeast if not using raw honey
- Room-temperature filtered or spring water
- 10–15 organic raisins
- Flavorful mushrooms of your choice, cleaned and diced. Suggested species include Shiitake, Chanterelle, Candy Cap, Maitake, and King Bolete

Equipment

- Cheesecloth
- Clean one-gallon glass jug
- Fermentation airlock
- Funnel
- Large metal spoon
- Large pot, ideally made of ceramic or glass
- Optionally, a fermenter's siphon
- Rubber cork for the jug that can accept a fermentation airlock
- Sanitizer sold for fermentation, such as Star San™

Recipe

1. Clean all equipment thoroughly with soap and water and then with the sanitizer following its included directions. Air-dry the equipment.

2. In the large pot, stir the honey into the water until it is dissolved. The proportions here are flexible, with one

part honey to four parts water being a good starting ratio (e.g. one to two pounds honey [or about one quart] to one gallon of water).

3. Actively stir the mix for five minutes, ensuring that the water is well aerated (which helps the yeast grow).

4. Cover the pot tightly with cheesecloth and place it in a 60–80°F (15–27°C) space that is out of direct sunlight. After several days, wild yeast from the air and honey will start to produce air bubbles in the mix, or *must*.

5. Clean, sanitize, and air-dry the jug, then transfer the must to the jug using your funnel.

6. Add the raisins to the jug, which help impart beneficial nutrients and tannins for the yeast.

7. Make a tea out of the mushrooms using the guidelines provided in Chapter III (page 83), then cool and filter the tea before adding no more than one cup of it to the mead, filling the jug to its brim.

8. Put the cork and airlock on the jug and place it back into the 60–80°F (15–27°C) space. (An airlock is a simple device that allows carbon dioxide (CO_2) to escape from the mead as the yeast convert the sugars into alcohol, while also blocking the entrance of insects into the container.)

9. After approximately 3–14 days, air bubbles will stop being released from the airlock, at which point the mix can now be carefully poured into bottles. If available, a fermenter's siphon is ideal for this stage as it will remove the mead from the jug while ensuring that the layer of dead yeast cells on the bottom of the jug (known as *lees*) does not make its way into your bottles, altering their flavor.

10. The mead can be consumed immediately, or left to age for months or years as its flavor becomes more refined.

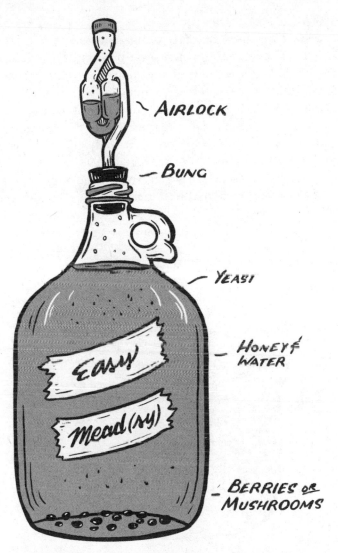

AIRLOCK

BUNG

YEAST

HONEY &
WATER

Easy

Mead (ry)

BERRIES OR
MUSHROOMS

*The essence of mead is made by mixing honey, water, and yeast in a clean container
and then adding an airlock to prevent insects from entering the container. From there,
variation is only limited by your imagination.*

Home Made Miso – GF / V

The miso paste used to make soups and sauces is produced through a multi-month fermentation process that is guided in large part by the mycelium of the mold *Aspergillus oryzae*. The spores of this fungus can often be found as "koji starter," though the easiest approach to making your own miso is to acquire some pre-grown koji from an Asian market. Koji is rice or barley that already has *A. oryzae* growing on it, and only needs to be mixed with cooked beans (soybeans are traditional, but chickpeas or barley can also be used), salt water, and a small amount of miso paste (which adds beneficial bacteria to the mix) to kick start a new batch of miso.

A double ferment with a rich history, miso is made by mixing fungi-infused koji rice, cooked beans, salt, and a bit of your last miso batch inside of a clean container, and then waiting several months or years.

There are a number of different miso types, all of which vary by the ingredients used, their ratios, and the amount of time the mix sits in a fermentation crock, with three months to two years being common time spans. Though you do have to wait a good while for this fungal ferment to mature, the result is a high quality, fresh, versatile, and umami-rich condiment that can easily find its way into many recipes when an abundance is on hand. As with other fungi-related hobbies, this one pays off in time and, though it may seem slow to get going, is easy to produce consistently once the equipment and materials are at hand.

OTHER FUNGAL FOOD FAVORITES
Mushroom Jerky - GF / P / V

If you come across a large haul of mushrooms while foraging, or if you grow more than you can eat, a popular method for preserving fruitbodies is to dehydrate and store them in sealed

A Shiitake harvest at MYCOLOGOS destined for jerky making.

jars. Dried mushrooms can be stored like this for years, and will readily rehydrate in a bit of warm water prior to cooking.

If you are feeling extra experimental, another option for making use of a bumper crop is to make mushroom jerky by blending the mushrooms with water, seasonings, and psyllium husks and then drying the resulting paste in a dehydrator to create mushroom "jerky." The ratio of ingredients will vary by the mushroom species used, their internal water content, and the amount of mushrooms you find, so make a few test batches to dial in your recipe. The ideal texture is like a thick jelly: spreadable yet malleable. Suggested flavors to add to your jerky mix include salt and pepper, tamari, miso, teriyaki sauce, and chipotle spice—though feel free to get as creative as you like!

Huitlacoche Tacos - GF / V

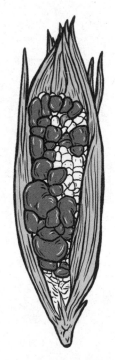

Huitlacoche (*Ustilago maydis*) is a fungus that infects corn kernels, transforming them into large purplish masses with a savory, umami-rich flavor and creamy texture. Though this fungal food has long been celebrated in Central America, corncobs infected by this fungus are commonly discarded by North American farmers, and so can often be salvaged from discard piles near corn fields. If you are able to source the spores of this species, you can even intentionally infect corn in your own garden. If you are not able to find fresh Huitlacoche, it is frequently found in the canned good sections of Hispanic markets. However you acquire some, cooking Huitlacoche is similar to preparing any other protein-rich taco filling: simply cook the fungus to the consistency desired (I like them a bit firm), adding your favorite seasonings as you go.

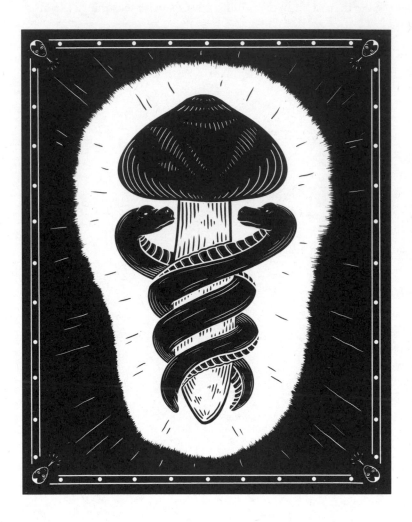

Chapter 3: GRAND CHEMISTS, GREAT HEALERS

*C*omplementing the nutrient density of edible fungi are the healing effects they provide for the mind and body. Fungi create hundreds of unique compounds that have been shown to enhance our ability to ward off disease, increase stamina, and manage imbalances in the body. In recent decades, research has proven that certain mushrooms and micro fungi provide a slew of health benefits for our body—a relationship as old as eating wild fruits (which are often covered in beneficial yeasts) or making plant medicines, which, we are increasingly finding, are filled with their own medicine-producing fungi. Centuries before the mold *Penicillium chrysogenum* was found to produce the antibiotic penicillin in 1928, herbalists and medicine people of India, Egypt, and Greece rubbed moldy bread on skin infections, having recognized the healing traits of these micro fungi. Likewise, the recent rise in medicinal mushroom awareness in the West is only the latest advancement of a practice that extends back thousands of years to some of the oldest records of the world's healing modalities.

It is in large part thanks to ancient written accounts of mushroom use in Traditional Chinese Medicine (TCM) that we even know of some of the most potent medicinal species being researched today. Perhaps not too surprisingly, as modern science catches up with these time-tested means of working with fungi, we are finding clear evidence that not only were these traditional healers well-versed in the medicines of their land base, but that their practices for making mushroom medicines still prove to be some of the best approaches to healing our bodies with fungi today.

At the same time, researchers are discovering that many of the medicinal compounds in fungi are found nowhere else in nature, and that a small number of species are exceptionally potent in their healing effect—with some mushrooms being revered for their impacts on chronic illnesses like lupus, dementia, and Crohn's disease. Along with these revelations has come a growing excitement around mushrooms in general, as well as the creation of new medicinal mushroom products that are more potent than the low-quality extracts that have been the industry standard in the West for decades. As I've watched the world of mycology grow in recent years, one of my greatest joys has been to meet people from around the world who attribute their personal health improvements to having access to the novel and potent medicinal mushroom extracts that have only been readily available since the modern mycocultural revolution began just a few short years ago.

YOU CAN GLUCAN

One of the most direct ways to partake in the healing benefits of mushrooms is by simply cooking and eating them. Unlike most table vegetables, which are more valued for their nutritional benefits than for other healing qualities, many gourmet mushrooms are not only delicious and nutritious, but also highly medicinal.

Though the exact effects vary by species, popular edibles like Shiitake (*Lentinula edodes*), Maitake (*Grifola frondosa*), Pearl Oyster (*Pleurotus ostreatus*), Lion's Mane (*Hericium erinaceus*), Nameko (*Pholiota nameko*), and Enoki (*Flammulina velutipes*) all host sugars in their cell walls that have been shown to benefit our immune system. These sugars are some of the same ones that comprise approximately 80–90% of the cell wall of mycelium (discussed in the Introduction), meaning that these mushrooms's fruitbodies are largely composed of beneficial compounds. Cooking mushrooms helps release these sugars and makes them more available to our body, as heat breaks the bonds in the chitin (the other 10–20% of the cell wall), which would otherwise make these sugars largely unavailable to our digestive tract.

When consumed, these "beta-D-glucan polysaccharides" (which are larger and more complex molecules than table or fruit sugars) seem to stimulate our immune system in a way that is akin to giving it a helpful workout, but without over-exerting it. Depending on the mushroom, this effect may help boost an otherwise compromised immune system, enabling the body to better fight off infections or disease, or it may soothe an overactive immune system that is prone to allergic reactions or autoimmune disorders.

For medicinal mushrooms that are too tough or woody to eat, making a tea or stock from the mushroom is an alternative method for extracting these sugars. By chopping a fresh or dried mushroom into small (approximately fingernail-sized) chunks, and then simmering those pieces for several hours in 160–175°F (70–80°C) water, these sugars will readily dissolve. Some species produce a sweet or bitter tea, while others can be bland, earthy, or have an overtly mushroomy taste. In addition, other nutrient-dense and simmer-able ingredients like seaweed, burdock, dandelion root, or animal bones, can be added to the mix, creating a potent broth to be savored immediately or frozen into ice cubes and administered as needed throughout the year.

EXTRA EXTRACTING

For a more potent extract, your mushrooms can also be soaked in high strength alcohol to collect their more heat-sensitive compounds (e.g. *monoterpenoids*, *sesquiterpenoids*, and *triterpenoids*). Many of these delicate compounds seem to have a soothing or anti-inflammatory effect on the body, but are lost or destroyed during the hot water extraction process described above.

In this approach, the mushroom is finely ground or chopped, placed in a dark-colored glass container, and then covered with 95% drinking alcohol (or, if that is not available, high-proof vodka) until the mushroom pieces are beneath two finger-widths of liquid. The jar is then placed in a warm area that is out of direct sunlight and shaken once a day for six weeks to help the alcohol extract these sensitive compounds. The mushroom pieces are then sieved out and set aside, while the liquid (or *tincture*) is temporarily placed in a clean, sealed jar.

To collect the mushroom's sugars, the next step is to simmer this same mushroom material in filtered or spring water for several hours. I like to use around fifteen times the mushroom's volume when measuring out the water (e.g. fifteen liters of water for every liter of mushroom material) and to simmer the mix at a low temperature (ideally no higher than 170°F [77°C]) until the water level has reduced to half its initial height in the pot. Once the tea has cooled, the mushroom material is filtered out and the water extract (or *decoction*) is mixed with the tincture at a rate of about thirty-percent tincture to seventy-percent decoction (i.e. thirty milliliters of alcohol extract for every seventy milliliters of water extract). Adding the tincture to the decoction (and not the other way around) is ideal as this avoids causing the water extract's sugars and oils to "crash" out of the solution (or become "un-dissolved"), which would make the final medicine cloudy. Once bottled, labeled, and stored out of direct sunlight, this "double extract" will retain its potency for a year or two, if not longer.

One of the benefits to making these extracts at home is that their concentration of medicinal compounds is often much higher than that of commercial products, which might appear to be of high quality, and yet are often only extracts of mycelium-covered rice (i.e. the ingredient list will note "myceliated rice" and not actual fruitbodies), and thus is really filled with more plant starch

1) MACERATE

D.I.Y.
Double
Extracts

3) SIMMER & REDUCE

TINC.

DEC.

2) STRAIN

30%

70%

4) STRAIN

5) COMBINE

Making high quality mushroom extracts starts with soaking finely chopped fruit bodies in high proof alcohol for six weeks. The liquid is then filtered out and the mushroom tissue simmered in water until a concentrated decoction is obtained. After filtering, the decoction is combined with the tincture to produce a shelf-stable double extract.

than potent mushroom compounds. I also prefer to make extracts from wild harvested mushrooms, for, as the wisdom of the elder herbalist goes, the medicine growing nearest to you is likely to be the best medicine for you.

DIY MORNING MUSHROOM BREWS

If you drink coffee, tea, or smoothies each morning, adding a bit of a mushroom extract to your daily beverage is an easy way to ensure you don't forget your mycomedicine. Adding a dropper-full of a double extract to your cup (after the drink cools a bit, if it's hot), or using a diluted mushroom tea in place of plain water when making your coffee are two easy ways to develop this habit. Another option is simmering down or dehydrating a mushroom tea or double extract to create a syrup for enlivening drinks. All

Processing medicinal mushroom at home starts with soaking (macerating) chopped mushrooms in high proof alcohol (left) and ends with a potent double extract that can support the mycofolk of your life.

of these options are likely to add a more potent effect than most commercial mushroom drink products, many of which don't detail the origin or potency of their fungal ingredients.

HEALED BY FIRE

Another option for engaging with the healing effects of fungi is to smolder a slow-burning species to experience the calming and cleansing effects of its smoke. Evidence of Indigenous peoples across the Americas, Europe, and Asia working with woody fungi in this way dates back thousands of years, with documented practices including the use of Amadou (*Fomes fomentarius*) by people in Siberia to cleanse the house of a deceased person, and the smoking of the Red Belted Conk (*Fomitopsis pinicola*) to alleviate headaches by the Northern Dene of Canada, among many other examples.

One of the most enduring practices with mushroom smoke is shared by several Indigenous groups of North America (e.g. the Inuit of Labrador and the Blackfoot of the North American Plains), who blend the ash of the Punk Mushroom (*Phellinus igniarius*) with tobacco, creating a blend known as *iqmik* ("things to put in the mouth") that potentiates nicotine's effect due to the high pH of the ash.

My favorite practice is to smolder Amadou mushrooms during special occasions when I want to cleanse the air and fill it with a uniquely fungal scent. I am particularly drawn to this mushroom for its connection to my own ancestry, as it has been found in ancient fire pits across Western Europe. Whenever I send a spark into its long, woody tubes and watch the ember slowly descend and spread across the conk's bottom, I sense a connection to forgotten traditions, and feel a bit more inspired to refine personal rituals for respecting my fungal ancestors.

HEALTHFUL SPECIES TO KNOW

If you are new to the world of medicinal mushrooms, the following seven species are some of the most highly regarded and commonly available as commercial extracts. All of them can be processed using the liquid extraction techniques noted above, though the amount you choose to consume will be based on your body's personal needs, which is best determined by a consultation with a health practitioner who is trained in medicinal mushroom administration.

Reishi [*Ganoderma lucidum*]

One of the most artistically depicted mushrooms of all time, Reishi has stood as a symbol for the elevation of one's life force that often comes from working with the woody fruitbodies of the wild. Its long-term consumption has been traditionally connected with increased longevity and of turning one into a "spirit being"—a potency reflected in its other common names: *Ling Zhi*, which is Chinese for "Tree of Life Mushroom," and *Mannentake*, which is Japanese for "10,000-Year Mushroom."

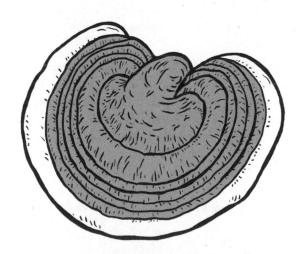

Researchers now know that this species produces more than 200 sugars and over 150 triterpenes— a potency that contributes to the strong bitter flavor of Reishi extracts. These compounds have shown the ability to lower cholesterol, reduce inflammation and high blood pressure, modulate the immune system, stimulate the libido, alleviate chronic pain, and support the liver, heart, and parasympathetic nervous system. Even the mushroom's spores have medicinal compounds that have been shown to prevent brain damage in a mouse model of Parkinson's disease. A potent mushroom, Reishi is not suggested to be taken at high doses regularly. But when properly worked with, its fiery energy readily reinforces the high respect it has earned over millennia.

Shiitake [*Lentinula edodes*]
The familiar, beautiful, and umami-rich Shiitake is one of the best-studied and most appreciated medicinal mushrooms. Not only is it high in vitamin D, this mushroom also produces a unique

amino acid (the building block of proteins) known as *eritadenine*, which may help reduce cholesterol. The fungus also produces compounds that are a mix of sugar and proteins, and which have positive effects against some types of cancer, while also acting as a *prebiotic* in our guts where they enhance the growth of beneficial bacteria, such as *Lactobacillus brevis* and *Bifidobacteria breve*, and may reduce the risk of colon cancer formation. Shiitake is perhaps best known for its production of the sugar

Shiitake mushrooms fruiting at the MYCOLOGOS Fungi Farm in Portland, Oregon.

lentinan, which has been shown to greatly support the immune system in individuals recovering from cancer treatments.

Maitake (*Grifola frondosa*)

Along with Reishi and Shiitake, Maitake (or, as it is also referred to in North America, Hen of the Woods) is considered one of the few *adaptogenic* mushrooms, meaning that it helps bring the body's systems into equilibrium. Much like the last two species, Maitake is highly regarded for its potential to combat various cancer cell lines and to support our body's immune response. Where this mushroom stands out is in its potential to support diabetics by simultaneously increasing insulin sensitivity, while reducing insulin resistance. Sufferers of the human immunodeficiency virus (HIV) might also benefit from taking Maitake extracts, which have shown notable effects against this virus in several studies. Likewise, the mushroom's ability to increase energy levels may be an important offering for people afflicted with chronic fatigue syndrome. In addition to all these properties, Maitake is also an *incredibly* delicious mushroom, especially when found in the wild.

Turkey Tail (*Trametes versicolor*)

Turkey Tail is one of my favorite mushrooms. Not only is it striking in appearance and easy to identify, it is also very easy to grow at home. Medicinally, the mushroom is considered one of the most important species of the

twentieth century as its unique compound krestin (a.k.a. PSK or Polysaccharide-K) was used for decades to boost the immune systems of people undergoing chemotherapy and radiation treatments for cancer. (In the 1990s, PSK use in these contexts was superseded by a more potent compound produced by an endophytic fungus that lives in the bark of the Pacific Yew tree.) In addition, Turkey Tail produces a sweet tea, which often fills my crockpot in the colder months. No matter how I engage with the mushroom, it seemingly calls out to be brought into the home and cherished for all of its colorful qualities. Really, what's not to love?

Cordyceps [*Ophiocordyceps sinensis* and *Cordyceps militaris*]

Found in high elevation Himalayan meadows of Nepal, Tibet, and China the "true" Cordyceps (*O. sinensis*) has long been revered for its energy enhancing effects on the body. Unlike the above species, which are decomposers, Cordyceps mushrooms grow out of the bodies of insects that the fungus had previously parasitized. Medicinally, the mushroom is regarded for its production of several compounds, most notably cordycepin, which can help strengthen the heart, lower cholesterol, increase stamina and libido, and even improve blood oxygen levels, making it helpful for athletes or people with low energy. However, as the mushroom has historically resisted artificial cultivation, its fruitbodies are currently wild-harvested to an increasingly unsustainable degree.

To counteract the cost and environmental impacts of working with *O. sinensis*, the home cultivation of the related and easy-to-grow species *Cordyceps militaris* (which also produces cordycepin and other medicinal compounds) has become popular among hobbyists in recent years. This trend might soon shift though as methods for growing *O. sinensis* fruitbodies have recently been discovered, opening the potential for this ancient and highly coveted medicine to inexpensively reach people around the world.

Almond Portobello [*Agaricus subrufescens*]

This almond-flavored and highly medicinal mushroom was commonly grown in North America around the turn of the twentieth century—often using practices similar to those used for the closely-related Portobello, Crimini, and White Button mushrooms (all of which are the same species, *Agaricus bisporus*). And yet, the alluring Almond Portobello is rarely grown in the West today. Throughout the last century, the mushroom has been

extensively studied for its ability to alleviate the symptoms of diabetes and to fight off viruses like poliovirus type 1 and hepatitis B and C. Its immune supporting effects are recognized as being so strong that nearly 500,000 people undergoing chemotherapy as a cancer treatment consume extracts of the mushroom each year in Japan. Additionally, the mushroom is also regarded as being a good antioxidant, a helpful anti-allergenic, and it is believed to provide liver support. Though not as easy to find in the grocery store, this mushroom does grow naturally in many parts of the world, and may even be a favorite at your local mushroom foraging club. Hopefully, as awareness around this species spreads, one outcome of the mycocultural revolution will be a return of this choice mushroom to grocery store shelves, replacing its less flavorful and significantly less medicinal relative.

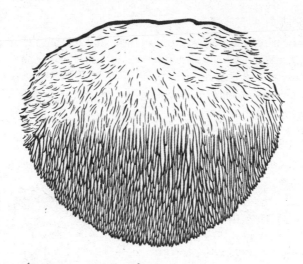

Lion's Mane [*Hericium erinaccus*]

This popular edible mushroom (also discussed in Chapter 2) has been elevated to the top shelf of notable medicinals in recent years due to the increased awareness of its unique *hericenone* and *erinacine* compounds. These triterpenoids are not celebrated for their capacity to help combat various cancer lines or support the immune system (traits that the mushroom also holds), but for their ability to increase the production of nerve growth factor (NGF) in our bodies. NGF is attributed to the healing and regeneration in our nervous system, and current research with Lion's Mane is looking into the mushroom's potential to help rebuild the "insulation" (or myelin) around our nerves, increase cognitive ability, repair neurological trauma and degradation, and potentially mitigate Alzheimer's disease and Parkinson's disease. With such powerful compounds, I believe that this mushroom will be among the most important of the twenty-first century. Learning to grow your own Lion's Mane and make potent medicine from its mycelium and fruitbodies are sure to be invaluable skills for anyone wishing to keep their wits in tip-top shape in the coming years.

THE PSYCHOACTIVE AFFECT

One of the most surprising fungal compounds to gain the attention of health researchers over the last two decades has been *psilocybin*: the famous tryptamine psychedelic found in approximately two-hundred "magic" mushroom species worldwide. Though psilocybin and similar compounds were investigated for their potential to treat psychological conditions in the 1950s (most notably alcoholism), after these substances were outlawed in 1965 research essentially stopped. Over the following decades, psychedelics became increasingly associated with the counterculture, and yet in the year 2000, researchers at John Hopkins University in Baltimore, Maryland were among the first in the world to regain government approval to study the potential health benefits of psilocybin on humans.

While the psilocybin used in these modern studies is synthetic and not truly derived from mushrooms, its reported effect is quite similar to the benefits that recreational psychedelics users describe coming from psilocybin-containing mushrooms. When one's mental and emotional state is properly prepared and

Psilocybe cubensis *is the quintessential psychoactive mushroom. Despite its effects being of moderate strength when compared to other psilocybin-containing mushrooms, it is the most commonly available species on the black market due to the ease with which it can be cultivated on a wide range of substrates.*

ready to handle the powerful (and potentially overwhelming) experience that psilocybin-containing mushrooms bring about, it's not uncommon for a person to have life-altering revelations about the negative ways that they treat themselves and others, while simultaneously being able to address suppressed memories that might underlie self-destructive habits.

In the synthetic psilocybin research of recent years, this "life review" effect has been used by therapists to guide study participants toward insights that might help them overcome an addiction, alleviate existential distress (such as that brought about by being diagnosed with a terminal illness like cancer), or reduce the intensity of otherwise treatment-resistant depression or obsessive-compulsive disorder. To date, results from these studies are not conclusive, yet do seem to suggest that, under controlled conditions, the clinical use of psilocybin may help select individuals suffering from such psychological afflictions.

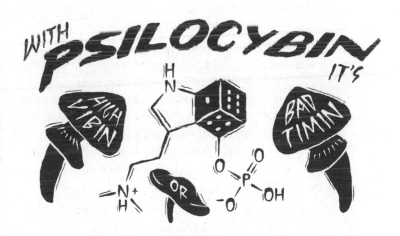

Whether an experience on psilocybin-containing mushrooms is beneficial or detrimental to a user is determined by a wide range of factors—a consideration that should not be taken lightly.

In addition, low doses of psilocybin-containing mushrooms are also regarded by people who suffer from debilitating (and often suicide-inducing) cluster headaches as the best—and often *only*—treatment for their condition. With these—and, likely in the future, other—beneficial effects of psilocybin becoming increasingly hard to ignore, I believe that this simple molecule will soon be made available to mental health practitioners in many parts of the world.

While this prospect is no doubt exciting, it is not without cause for concern. Being such an effective means for rewiring the brain, psilocybin consumption is inherently risky, as the history of psychoactive mushroom use has shown. Many inexperienced users of the drug report entering an intense and often frightening state of dissociation during the multi-hour psilocybin experience—trips so bad that their effect can leave a long-term psychological scar. I have personally known some individuals who suffered mental instability for years after such experiences—including one acquaintance from my teenage years who never fully recovered from a psychosis-like mushroom trip at the age of 16.

Though the protocols being developed around psilocybin administration aim to reduce the rate of these negative effects, the risk remains. This is especially true in the U.S., where in recent years psilocybin has been decriminalized in several cities, effectively making private possession of the substance akin to a traffic violation in those areas. While these changes do, thankfully, get us closer toward eroding the costly and ineffective war on drugs, they also increase the risk of more people having a devastating time on psilocybin.

In addition to these effects on the individual, we must also wonder how increased consumption of psychoactive substances could impact society. Could increased psilocybin use lead toward greater peace between individuals and nations, as some of its advocates suggest? Or could the drug's consumption increase pacification among the masses, and thereby reduce the amount of civil unrest that leads to more systemic changes in society?

While I don't claim to know the answers to these questions, I do believe that the seemingly inevitable shift toward increased psilocybin decriminalization—if not complete legalization—in the U.S. and abroad will be a major turning point in history and may even mark the start of a new era of human-fungal relations. What direction that turn takes, though, is ultimately up to all of us invested in finishing the current chapter in the human-fungi story on a good note, while inspiring the next generation of mycophiles to carry the torch where we left off.

Spawning the Present

The mycocultural landscape of the twenty-first century has grown through several evolutionary leaps in just two short decades. With the advent of the internet, information on fungi has rapidly spread from a small number of universities and mushroom farms—where it had been tightly guarded throughout the prior century—to the kitchens and conversations of people around the world. This growing awareness has removed the outdated notion that mushrooms and molds are nothing more than odd and dangerous, while opening space for anyone to get involved in the new culture's uprising. The democratization of mycology has enabled fungal allies who'd long struggled to learn the mycological arts from textbooks and industrial manuals to join a global community of mycophiles spreading their knowledge through online and in-person gatherings. As newcomers to the science meet welcoming veterans of its many trades, the once tenuous world of mycology now finds itself saved from a burial on library bookshelves and instead constantly refreshed within each new context it's given.

A shared sense of values and intentions is essential for the sustainment of any culture. And so we find that the many-branched world of do-it-yourself and do-it-together mycology has begun developing guiding principles to ensure its growth remains healthy and exponential. In the ever-evolving mushroom cultivation sphere, emphasis is increasingly placed on growers openly sharing improvements on the art's various techniques, while giving credit to a method's originator and other adapters. Maintaining this shared respect is critical for ensuring that newcomers recognize that their "fresh eyed" insights are greatly needed and will be welcomed by more experienced growers, so that all may benefit. Though mushroom farming has historically been an industry wrought with trade secrets, in the mycocultural revolution the potential for cultivation to alleviate many social and environmental issues, such as job shortages and finite resource management, has replaced the old veil of distrust amongst growers with a shared vision of a more resilient and fungi-filled future.

In the community of mushroom medicine makers, the bar continues to rise regarding the quality of products produced by small and large companies and the information that is used to sell them. Whereas the industry has long relied on the limited understanding of customers to sell low potency mycelium-based products in place of more vital (and expensive to produce) whole mushroom extracts, today's better-informed community of healers and patients is demanding higher quality medicines and more honest marketing from providers. As Western extracts come to more closely match the potency found in many Asian products, the great potential for fungi to support individuals and communities afflicted by chronic disease becomes increasingly tangible and central to advancements in traditional and alternative therapies.

In the Radical Mycology movement, we focus on creating physical, digital, and artistic spaces where everyone is encouraged to teach and inspire others about their unique perspectives on the state of mycology. Building a community between fungi-lovers of diverse cultural, political, and spiritual backgrounds has been a cornerstone of our work since the first Radical Mycology Convergence, as we believe that it will only be through an open dialogue on what it means for the human and fungal realms to align that new paths can be grown out of modern life's limits.

I see all of these advancements as only starting places, however. To ensure the revolution continues to grow, the more intangible aspects of a self-sustaining and inspiring mycoculture—its values and philosophy—must be further refined. More than ever, artists, poets, and creatives of all mediums and styles are needed to help translate the practical aspects of applied mycology into enduring expressions that transcend cultural barriers. We need a plurality of voices in the revolution, so that we each discover the dialogue with and about fungi most closely aligned with our own worldview.

Compared to nearly any other science or area of interest, the innumerable forms of fungi have been poorly represented in films, music, architecture, literature, sculpture, and paintings throughout human history. Our symbolic language has been mycologically stunted, leaving our capacity to envision their potential shortsighted and perpetually topical. As we move deeper into the revolution, the practical talents of inventive cultivators and the mythos crafted by visionary mycofolk are providing the needed frameworks for infusing fungi throughout society and lowering the learning curve for those new to the topic.

At the same time, early adopters and mycoambassadors are invited to help build a more participatory mycoculture, wherein a shared passion for all that fungi offer sets aside personal differences and ensures a collaborative effort is held in the dissemination of mycology across subcultures and smaller spheres of influence.

Wherever you stand on the spectrum of myconoob to fungi fanatic, your interest in mycology is in itself a powerful means for shifting any social or environmental paradigm toward a more fungi-conscious perspective. For as we each spread our personal skill set for working with fungi throughout our closest communities, we collectively share in a silent shift from an old world lined with fungi fears into one renewed through the many gifts of mycology.

Chapter 4: GROW YOUR OWN MUSHROOMS, FOR EVERYBODY!

*I*n the thick of the revolution, we now realize that one of the first steps in building the most resilient, fungi-inspired society is growing a larger community of mushroom cultivators to expand the mycelial networks that might transform our waste streams into food and soil, just as they do in fields and forests. Such a potential is now possible thanks to the recent simplification of once-intimidating cultivation methods—an unprecedented leap forward in the history of human-fungal relations that comes in the wake of centuries of constant improvement by innovative cultivators on both the large and small scale.

The benefits of promoting and sharing the gifts of cultivation with our communities are many. Not only does the work provide nutrient-dense whole foods and potent natural medicines to the grower and those they share their crops with, but also job opportunities for people of all social and economic backgrounds. Unlike plant growing, mushrooms can be cultivated with minimal lighting and energy inputs, while readily taking advantage of vertical space. Simultaneously, mushroom growing makes direct use of common organic residues, such as wheat straw, manure, and coffee grounds, enabling local economies to ensure greater food security with limited resources. Mushrooms can be dried, stored, powdered, or processed into medicines or food flavorings, while their mycelium can be applied as a soil amendment or grown into furniture or other functional objects, reducing our use of plastics and other building materials. These are just some of the applications that growing mycelium offers, with undoubtedly more to be recognized in the coming years as cultivating mushrooms becomes as commonplace as vegetable gardening.

People of all ages are encouraged to join the cultivator community's exploration of this untapped potential as we work together to build a healthier, fungi-filled future. With the basic questions of how to grow mushrooms now largely resolved, there has never been a better time to learn this vital life skill and enjoy the many gifts of working with fungi simply by following the methods described in this chapter. Together, the world's cultivators become the mycelial network of the human world, bridging the varied spheres of society through the powers of fungi that gracefully transmute and move the elements of life, just as they transform and motivate those who ally with them.

WORKING WITH FUNGI, THE EASY WAY

Fermenting with microbes to produce food or medicine follows the same general principle, regardless of the final product: with the right combination of organism, food, and shelter, patience is the final ingredient needed to obtain a good yield. For sauerkraut, all that is required to start a batch is shredded cabbage, salt, water, a clean jar, and a not-too-warm space, while crafting a simple wine calls for fruit juice, water, yeast, and a sanitized jug. Mushroom growing follows a similar pattern, in that mycelium is fed a mixture of common ingredients and then placed in a clean container at the right temperature to ferment that substrate for several weeks before fruiting mushrooms from it.

Where mushroom cultivation differs from other fermentation skills is in its greater need for clean tools and materials, and in the greater amount of materials and labor involved—a result of the larger size of mushrooms when compared to the yeasts in wine or microbes in sauerkraut. The outcome of these two differences is the mushroom grower's need for specialized equipment that is not used for other types of ferments, but which can be made inexpensively and reused for many years.

Regardless of the technique used to grow mushrooms (of which there are *many*), I think of the whole process in four key stages. The first three expand a small amount of mycelium obtained from spores or a tissue sample into a larger volume through a series of feeding steps, while the fourth stage sees that bulked-up mycelium either transformed into mushrooms or applied in an outdoor installation.

Traditionally, the first two stages were the most difficult of the whole process as they called for the use of lab-based tools and techniques to isolate pure mycelium that was free of bacterial and mold contaminants. Today, these more demanding stages are avoided by most growers who opt to purchase clean starter mycelium, known as *spawn*, for their cultivation work rather than struggle to grow their own. In the

Growing mushrooms is similar in principle to other fermentation practices. While the grower tries their best to provide food, water, air, and a clean environment, the microbes work their magic to transform the ingredients into a novel delight.

Mushroom cultivation starts with isolating pure mycelium. That culture is then exponentially amplified on sterilized grains, which is then applied to a fruiting substrate (in this example, wheat straw). After an incubation period, fruiting is initiated and soon a harvest is celebrated.

methods below, this is the route I suggest if you are new to mushroom cultivation. For detailed methods on producing your own spawn and engaging with all four stages of mushroom growing, see *Radical Mycology: A Treatise on Seeing and Working With Fungi*.

DESIGN(AT)ING WORKING SPACES
The first step in the cultivator's rhythm is designing a functional workspace that best matches your goals, budget, and time

availability. At a minimum I suggest designating one area for tools, another for storing, mixing, and cooking ingredients, and a humid environment for fruiting your mushrooms. The first two of these spaces can fit in a cupboard or closet near your kitchen, or organized on a shelf in the corner of a garage, basement, or shed. The fruiting environment would ideally be situated in the same area—so that everything stays close at hand—but it might need to be located elsewhere to accommodate for its specific requirements, as described below.

If you are new to mushroom growing, starting small with all of these spaces will keep your practice more manageable and, ultimately, more enjoyable. A common challenge for new mushroom growers is soothing the desire to practice every

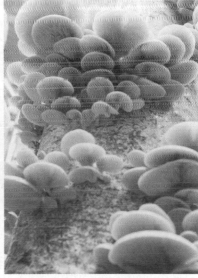

Scenes from the MYCOLOGOS Fungi Farm: (Left) Reishi mushroom mycelium incubating. (Right) Oyster mushrooms fruiting from a tube of pasteurized wheat straw (a "strawsage").

cultivation method simultaneously and to acquire all of the tools that each one calls for—with the result almost always being an inability to keep up with the quick-growing mycelium. While I can relate to this feeling, I suggest keeping things as simple as possible and practicing one method at a time when starting out, so that you not only have greater success, but also reduced stress and increased levity in your work with fungi—the foundation for any life-long relationship.

MISTS OF A FRUITING ENVIRON

Though it may seem like putting the spore before the mushroom, I suggest building your humid fruiting environment before any mycelium is at hand. This ensures that once your mushrooms are ready to fruit there will be no delay in starting the fruiting process, a setback that can cause a crop to get stunted, reducing yields. Several fruiting environment designs are described below, all of which should ideally be sited in a location that is clean, regularly visited, not lit by direct sunlight, cool in temperature, and ventilated with fresh air. A kitchen counter that is shaded and not too hot, or a clean section of a cool basement or garage are good sites for a fruiting environment, whereas closets with low air flow, the warm tops of bookshelves, or porches with varying temperature and humidity levels are not preferred.

The best sites have bright, full spectrum lighting, which tends to produce deeper colors, richer flavors, and higher concentrations of medicinal compounds in mushrooms. Indirect sunlight or LED light bars or ropes with a temperature of 6000–6500 kelvin (K) are often used to match this need.

Likewise, many mushrooms grown at lower temperatures tend to be more flavorful, with some species requiring the air to be below 65°F (18°C) to trigger their mycelium into producing fruitbodies. Placing the mushrooms in a cool part of the house is a passive way to obtain a lower temperature without using extra electricity on an air conditioning system. Alternately, if you live in a warm part of the world, you may wish to only grow mushrooms that prefer higher temperatures, such as Pink Oysters (*Pleurotus djamor*) or Paddy Straw Mushrooms (*Volvariella volvacea*), and avoid the need for cold air altogether.

The most important element of a fruiting environment is its air quality, which should be high in oxygen and relative humidity. As mushrooms mature, they expel carbon dioxide (CO_2) and lose moisture off of their surface through evaporation (just like we do). To ensure that the mushrooms receive the oxygen their cells require for respiration, the CO_2 in the fruiting environment needs to be exchanged with fresh air several times an hour. At the same time, the air needs to stay humid (typically at around 85% relative humidity) so that the mushrooms don't dry out and stop growing. Balancing these two air quality factors is the trickiest aspect of designing a good fruiting environment, and one that will likely take the most time when first dialing in your space. Once this

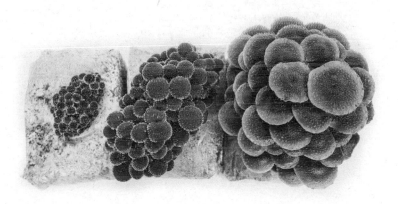

Once fruiting begins, most gourmet mushrooms mature within just a matter of days. Harvesting is generally best done before the mushroom's caps flatten out.

balance is achieved, though, a well-maintained fruiting space will enable you to grow fresh mushrooms all year, making that initial time spent well worth the effort involved.

TWO SIMPLE FRUITING SPACES

To account for the above considerations, several inexpensive home-scale fruiting environments have been designed by the cultivator community in the last two decades that remain popular due to their low-tech approach.

The simplest of these is to drape a clear plastic bag over a rectangular frame made of wire coat hangers, cutting small (e.g. 0.125–0.25-inch wide) holes or slits in the bag to provide air flow, and then misting the inside of the bag daily to keep its internal humidity level high.

Another approach is to drill 0.25-inch (0.6 cm) holes across the four walls of a plastic tote (the larger the better, so as to house

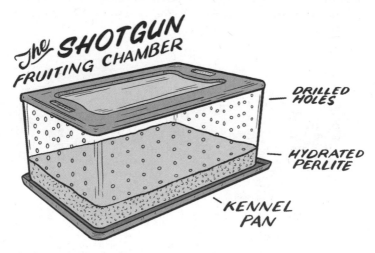

The shotgun fruiting chamber is easily made by drilling small holes around a plastic tote, then filling the bottom five inches with hydrated volcanic rock or perlite.

more growing containers), each spaced 2 inches (5 cm) apart in all directions, creating a "Shotgun Fruiting Chamber" (SGFC). The bottom of the SGFC should also have several drainage holes drilled in it before being covered in 5 inches (12 cm) of volcanic rock pebbles that have been soaked in filtered water. With the tote's lid in place, the holes will provide for passive air exchange, while the majority of the water vapor released from the volcanic rock will be retained by the walls, keeping the internal humidity levels high. If needed, the inside walls of the SGFC may need to be misted with cold filtered water occasionally to maintain a light fog on their surface—the visual cue of a proper SGFC humidity level.

While these systems are great for getting started, their biggest drawbacks are in their limited holding capacity and their need for almost-daily attention. If you'd prefer a larger or more automated system, see *Radical Mycology: A Treatise on Seeing and Working With Fungi* for information on designing automated fruiting environments that incorporate timer-controlled lights, fans, and humidification systems.

GET GROWING!

With your workspaces squared away, it's time to get growing! Below are the first projects I suggest working through as you begin your exploration of the wide and varied world of mushroom cultivation. Time tested and easy to practice, these cornerstone methods will not only enable you to grow mushrooms nearly anywhere and on nearly any budget, but also demonstrate the essential techniques of indoor mushroom growing. Likewise, the experience and confidence gained through the practice of these protocols will lay a foundation for later developing your own approach to cultivation and for designing experimental trials— one of the art's most exciting aspects.

*Chestnut mushrooms (**Pholiota adiposa**) fruiting at the MYCOLOGOS Fungi Farm.*

A highly-branching Reishi variety fruiting in a home-scale automated fruiting environment.

Fruiting Ready-To-Grow Kits

The easiest means for growing fresh mushrooms at home is to acquire a ready-to-fruit grow kit from another cultivator. Depending on where you live, this may be available at your local farmer's market or from a number of online retailers. Once obtained, all that is needed is to open the container that the mushroom's mycelium is growing in (typically by cutting slits in the bag it's growing inside of) and then placing the container in your fruiting space after ensuring that the kit's temperature and air quality needs will be provided. One to three weeks later, mushrooms should begin to appear, with most species being ready to harvest three to ten days later.

Most mushroom farms discard their mushroom grow kits after they have fruited once, despite the fact that most kits will produce more mushrooms weeks later. Collecting these "spent" blocks from your local grower and then fruiting them at home is a low-cost way to grow an abundance of fresh fruitbodies.

Tried-n-True Oysters on Straw

The Pearl Oyster mushroom (*Pleurotus ostreatus*) and its close relatives the Phoenix Oyster (*P. pulmonarius*) and Pink Oyster (*P. salmoneostramineus* and *P. djamor*) are among the most commonly cultivated gourmet mushrooms in the world. Not only are these species rich in flavor and healthy to eat, they are also highly tolerant of less-than-ideal growing conditions, making them good species for both beginner and experienced growers. These mushrooms are also well known to grow on dozens of common agricultural residues, such as corn cobs, peanut shells, and hemp stalks, making them essential players in the creation of local, resource-conscious food streams that use materials which might otherwise be burned in a field.

*Pearl Oysters (**Pleurotus ostreatus***) and their close relatives are the best species for first time growers to work with. Tolerant of dirty work flows and less-than-ideal fruiting environments, these forgiving species fruit in large clusters from a wide variety of substrates.*

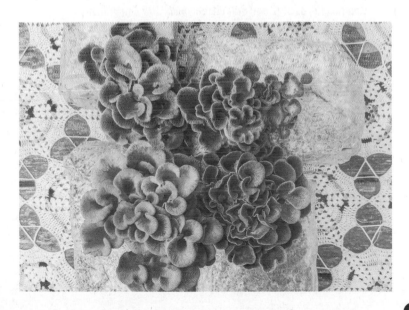

Among these potential substrates, the most commonly worked with in North America is wheat straw, due to its commonality, low cost, and ease of preparation. After decades of growing Oysters on straw, several approaches to preparing this material have been developed, with one of the most popular being to pasteurize shredded straw in hot water. After the straw is cooked for an hour, drained, and cooled, it is then mixed with Oyster *grain spawn* (i.e. Oyster mycelium growing on rye berries, sorghum, or millet) in a plastic bucket or bag to incubate for several weeks before being placed in a fruiting environment.

Materials

- **Cleanable work surface** – A metal or plastic tabletop is ideal, though a clean tarp or plastic sheet laid on the ground or on a wooden table will also work.

- **Clean, heavy weight**

- **Electric drill, 0.75" (2 cm) drill bit, and 0.25" (0.7 cm) drill bit**

- **Food-grade 3- or 5-gallon buckets**

- **Heat source** – Stovetops that can support a heavy weight or a large propane burner are popular options. If you live in a sunny part of the world, off-grid solar water heaters can readily achieve the temperature needed for pasteurizing straw.

- **Large metal container**

- **Oyster grain spawn** – This is best acquired from a reputable spawn provider in your part of the world. Many strains are available, with each having their own fruiting pattern (e.g. large clusters of small mushrooms vs. clumps of several large mushrooms), coloration (e.g. white, blue, or gray), and preferred fruiting temperature window. If possible, select two or more strains that appeal to you to compare their differing habits and traits.

- **Porous, heavy duty sack made from natural fibers** – Old pillow cases or burlap bags like those used to ship organic coffee beans both work well.

- **Spray bottle** filled with **70% isopropyl alcohol**

- **Thermometer**

- **Water** – The best water is from a spring or clean well, with no additives applied. If you are using tap water, let a non-lidded pot of the water sit out for 24 hours to allow its chlorine gas to evaporate out.

- **Wheat straw** – Organic straw is preferred, so as to avoid fungicides. Be sure to get straw and not hay, as the latter has seeds that can cause contaminant growth.

- **A means to shred the straw** – Common methods include using a yard debris shredder, running a lawn mower over the straw on a concrete slab, or running a lawn edger inside of a trash can filled with the straw.

Method

1. Drill 15 equally-spaced 0.75" (2 cm) holes around each bucket and five 0.25" (0.7 cm) drainage holes in the bucket bottoms.

2. Fill your metal container two-thirds full with water and warm it to 160–170°F (70–77°C) with your heat source.

3. Shred the straw into 1–3" (2.5–7.5 cm) sections (or as close to that as possible) using one of the methods listed above, then dump the material into your fabric sack.

4. Submerge the sack in the hot water, then place the weight on top of it.

5. Keep the straw in the water for an hour, ensuring that the temperature of the straw stays between 140–170°F (60–77°C) by turning the heat source on and off as needed.

6. Remove the straw bag and set it to drip and cool off for several hours. (If desired, the remaining water can be used to treat a second batch of straw, but not a third. After the second batch, the water becomes too nutrient rich and is likely to encourage contaminant growth on the straw.)

7. Once the straw has stopped dripping and is cool, spray the worktable with isopropyl alcohol, wipe the insides of the buckets and their lids with alcohol, and break up the Oyster grain spawn by gently crumbling it inside of the bag that it came in.

8. Wash your hands and forearms with soap and water and then spray them with isopropyl alcohol.

9. Sprinkle several inches of straw into the bucket, gently breaking up any clumps as you go.

10. Sprinkle half a handful of grain spawn on top of the straw and then add a couple more handfuls of straw.

11. Repeat steps 9 and 10 until the bucket is half full, then use a clean bucket to press the material down firmly.

12. Continue filling the bucket and packing the mixture until the bucket is full, then clean and secure the bucket's lid.

13. Label the bucket with the date and mushroom species worked with, and then place the bucket in a 65–75°F (18–24°C) space where it will not be disturbed.

14. Once small mushrooms begin to develop from the holes in the bucket (typically 2–4 weeks later), move the bucket to your fruiting environment or drape a large, perforated

plastic bag over it, ensuring that the bag is large enough that the plastic will not touch the growing mushrooms.

15. After the first crop, or *flush*, is produced, the bucket can stay in the fruiting environment, where it may produce a second flush 1–3 weeks later.

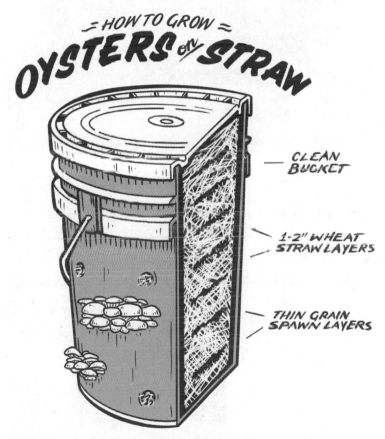

= HOW TO GROW =
OYSTERS on STRAW

CLEAN BUCKET

1-2" WHEAT STRAW LAYERS

THIN GRAIN SPAWN LAYERS

Growing Oyster mushrooms on pasteurized wheat straw is an essential skill for any grower. After drilling and cleaning the container, fill it with alternating layers of straw and grain spawn before packing it all tightly and securing the bucket's lid.

If you're making multiple straw buckets, lining them up and filling them all at once shortens work times, as does using a clean bucket to pack the contents down firmly.

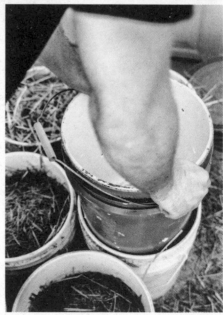

Technique Variations

- **Lime-treated straw** – Soaking shredded wheat straw in a high-pH water bath for 8–12 hours is an alternative low-tech treatment method that is popular, even though it tends to produce a lower yield than heat-treated straw. Adding 0.5 cups (120 mL) of hydrated or agricultural lime (readily available at most plant nurseries) to every 14–16 gallons (50–60 L) of water is typically all that is needed to create the bath, though be sure to get some pH strips to ensure you are at the 12 pH or higher. After soaking, the straw is drained and inoculated as described above. Unlike the water used in the hot soaking method, lime water can be used multiple times, so long as the pH level remains above 12.

- **Cold fermented straw** – The most low-tech method for treating straw is to ferment the substrate in stagnant, non-chlorinated water for 7–14 days. Once the straw and water smells sour and pungent, the straw is lifted out, drained, and inoculated as described above. This approach generally is the lowest yielding treatment method, but is worth experimenting with at least once, if only for the nasty smell alone!

- **Oysters on coffee and cardboard** – One or two days old coffee grounds are an increasingly popular substrate for Oyster mushrooms as they are typically free and abundant at coffee shops. Coffee grounds don't need any pre-treatment, but should be drained enough that they don't create standing water in the growing container. Using the grounds as a substrate is essentially the same as with straw, though the grounds should not be packed down after the bucket is full as this will cut off airflow to the mycelium. As an extra step, strips of tape-free corrugated cardboard that have been soaked and drained can be added between each one-inch layer of coffee grounds to add an additional carbon source and increase air flow in the substrate.

- **Fresh Oysters as spawn** – Because Oyster mushrooms grow so vigorously, fresh mushrooms from the grocery store may work as a mycelium source for your substrate. (Remember, a mushroom is just structured mycelium.) The first time I grew Oysters on coffee was by burying torn pieces of several fresh Oysters into a bucket full of fresh coffee grounds and then loosely applying the lid. After forgetting about the bucket for several weeks, I opened it up just in time to find it bursting with dense mycelium and the start of mini mushrooms that quickly matured. This method doesn't always work, but is pretty rad when it does!

Growing Reishi, Easily

Compared to the wide appetite of the Pearl Oyster and its allies, most gourmet and medicinal mushrooms are picky eaters that need to grow on hardwood sawdust before fruiting. Additionally, cultivators have discovered that supplementing this sawdust with minerals and proteins will greatly increase yields, albeit at the added cost and complexity of sterilizing the whole mixture to ward off contaminating microbes brought on by the supplements.

For small-scale growers, an alternative approach to working with sterilized supplemented sawdust is to inoculate plain, non-supplemented sawdust with a higher-than-normal amount of grain spawn, so that the grains themselves provide the additional nutrients. Though cost prohibitive at a large scale, this method is an ideal, low-stress method for getting started with wood-loving species.

One method I suggest is mixing hydrated hardwood sawdust pellets with grain spawn inside of a self-contained fruiting environment referred to as a monotub. Under optimum

conditions, the mushroom's mycelium will run through the sawdust and, several weeks later, fruit inside the monotub without any maintenance. This set-and-forget system is excellent for growing many types of vertical-growing mushrooms (as opposed to those that form horizontal fruitbodies) as monotubs can be stacked, making use of vertical space, or tucked into an unused corner of a room.

Making a Monotub

Materials

- **Plastic tote** – The size of the monotub is based on the amount of substrate you will use.
- **Black plastic trash bag**
- **Electric drill**
- **One-inch (2.5 cm) hole saw**
- **Micropore (first aid) tape**

Method

1. Use the hole saw to drill several holes on the long sides of the tote: 2–3 holes 5 inches (12.5 cm) from the bottom of the tote and 1–2 holes 3 inches (7.5 cm) from the top of the tote. (Larger totes require more holes, so as to increase air circulation.)

2. Line the bottom five inches of the tote with the black plastic bag, trim off any excess bag material, then tape the bag material to the tote's walls.

3. Clean the inside of the tote and its lid with isopropyl alcohol.

4. Cover the outside of the holes with micropore tape.

Materials

- **Clean monotub**

- **Food grade hardwood sawdust pellets** – Often found at hardware stores, these pellets are used for smoking meat and, thanks to their manufacturing process, are "clean enough" out of the bag that they don't need much processing for use as a substrate.

- **Large measuring device**

- **Large spoon**

- **Micropore (a.k.a. first aid) tape**

- **Mushroom grain spawn** – Any species that is known to fruit on sawdust and develop a vertical fruitbody can be trialed with this method. Suggested species to experiment with are Enoki, Pioppino, Reishi, Pearl Oyster, and King Oyster.

- **Polyfil**

- **Spray bottle** filled with **70% isopropyl alcohol**

- **Two large pots**

Method

1. Decide on the amount of grain spawn you are able to acquire to then determine the size of your monotub. For every part grain spawn, three to four parts hydrated sawdust will be needed. Add the spawn volume and sawdust volume to determine your total substrate-spawn volume. Convert this volume to cubic inches, then divide this number by 4 inches (10 cm) (the height of the mixture, once it is in the monotub) to determine the surface area (footprint) of the monotub needed.

2. Determine the amount of pellets needed by multiplying the volume of hydrated sawdust determined in step 1 by 0.75. Measure out this volume of pellets and set them aside.

3. Prepare a monotub of the appropriate size following the instructions in the text box above.

4. Fill one large pot with water and set it to boil.

5. Clean the other large pot with soap and water and then spray it with isopropyl alcohol. Fill the pot no more than half full with the sawdust pellets, then add roughly half as much boiling water to the pellets (e.g. for one gallon of pellets, add half a gallon of boiling water). The heat of the water will help clean the pellets as they quickly hydrate over the following minutes.

6. Once the water has absorbed into the pellets, clean the large spoon with isopropyl alcohol and use it to crumble any intact pellets and mix the water and sawdust evenly.

7. Use a clean (i.e. washed and alcohol-sprayed) hand to check the sawdust's moisture content. The sawdust is properly hydrated when it feels moist but not overly wet (much like a wrung out sponge). It will also stick together slightly when squeezed, while releasing only a drop or two when squeezed firmly. If the sawdust is too dry, add more boiling water.

8. Once the sawdust is cool to the touch (or around 75–80°F [24–27°C]), clean the measuring device with isopropyl alcohol, allow it to air dry, then measure out the volume of hydrated sawdust determined in step 1, moving the material into the monotub as you measure it.

9. Break apart the grain spawn by gently crumbling it inside of the bag that it comes in and pour it into the monotub.

10. Use the clean spoon to mix the sawdust and spawn evenly.

11. Secure the clean lid of the monotub and label it with the date and mushroom species before placing it in a 65–75°F (18–24°C) space where it will not be disturbed.

12. In several weeks, the mycelium will run through the sawdust. During this time, avoid the desire to open the monotub's lid as this will cause a drop in humidity that can ruin the crop.

13. Once mushrooms start to develop inside the monotub, quickly remove the micropore tape from the tub's one-inch holes and stuff them lightly with polyfil, providing a fresh air supply to the maturing mushrooms.

14. If you are growing Reishi, the mushroom will likely grow up the walls of the monotub or form "antlers" along the surface of the substrate. These fruitbodies should be harvested after their tips stop producing a white edge (signifying growth has ceased). If you are growing one of the other suggested species, harvest the mushrooms before their caps have flattened out. If luck is on your side, the monotub may produce another crop in several weeks, especially if you mist the inside walls of the tote with water before closing it back up.

The combination of an incubation and fruiting environment, monotubs are a popular approach to growing mushrooms. After inoculating several inches of pasteurized substrate with grain spawn, the tub's lid is secured and air filters are applied to all holes. Mushrooms fruit within several weeks, with no maintenance required in the interim.

Portobellos x Horse Manure

Animal waste is one of the oldest mushroom growing substrates in the West, and also one of the trickiest to work with. Not only must the material be sourced from a stable or ranch, it also needs to be extensively handled and processed to yield the best results. Still, devout growers will jump over these additional hurdles to grow the small number of *Agaricus* and *Psilocybe* species that prefer this substrate. For the following recipe, I have not included exact measurements for ingredients, but instead ratios. This way, you can choose to do a small or large batch, depending on your budget, while still following the same instructions.

Materials

- **3 large pots**
- **Aluminum foil**
- **Clean weight**
- **Food thermometer**
- **Freshly used coffee grounds**
- **Hand towels**
- **Horse manure** – Ideally, the manure has been rained on and dried in the field, so that any residual urine (which can impact mycelium growth) has been removed. Cow manure can also work, but the fibrous quality of horse manure makes it an ideal substrate over most other domesticated animal wastes.
- **Knife**
- **Mushroom grain spawn** – The method below will work with most species that grow on compost or manure in the wild. Grain spawn for Portobello and White Button species is easy to acquire from spawn producers around the world,

while *Psilocybe* spawn is only available in regions where the cultivation of these species is legal.

- **Shallow tray** – A baking tray or similar shaped plastic tray are good smaller options, while shallow totes designed to store items under beds are a larger option.

- **Spray bottle** filled with **70% isopropyl alcohol**

- **Sturdy gloves**

- **Vermiculite, peat moss, agricultural ("hydrated") lime,** and **calcium carbonate** – All of these ingredients are commonly available at plant nurseries

- **Wide mouth canning jars**

Method

1. Wearing gloves, crumble the horse manure as finely as possible inside one of the large pots. Slowly incorporate water until the manure does not drip when held in the hand, but does produce a small number of drops when lightly squeezed and a short stream of water when firmly squeezed.

2. In a separate pot, hydrate the vermiculite to a similar water holding capacity as the manure.

3. Mix the materials in a ratio of one part vermiculite to two parts manure, then add 1/8 part fresh coffee grounds. Keep the substrates fluffy as you mix them together— avoid squeezing out air from the mix.

4. Loosely fill the canning jars with the substrate and cover them with aluminum foil.

5. Line the bottom of the third large pot with hand towels, fill the pot with the substrate jars, then fill the pot with

cold water until the jars are two-thirds of their height in water. If the jars float, cover them with a weight.

6. Insert the thermometer into the center jar (through the foil), ensuring the end of the thermometer does not touch the bottom of the jar.

7. Turn on the heat source and keep an eye on the thermometer. Once the gauge hits 130°F (55°C), turn off the heat. The temperature of the substrate should continue to climb and stabilize around 150°F (65°C). As needed, heat the pot to maintain a temperature of 140–170°F (60–77°C) in the jars for one hour.

8. Remove the jars and allow them to cool.

9. Wash the shallow tray with soap and water and spray it with isopropyl alcohol.

10. Break up the grain spawn by gently crumbling it inside of the bag that it came in.

11. Once cooled, sprinkle a 1" (2.5 cm) layer of manure in the tray. Add a thin "salting" of grain spawn on top of the manure, such that there is one kernel roughly every centimeter.

12. Continue layering the substrate and spawn until the tray is filled to one inch below its brim.

13. Cover the tray with foil and poke a small hole in the foil every 3 inches (7.5 cm). Label the tray with the date and species then place it in an area where it won't be disturbed.

14. Once the mycelium is run through the substrate and visible on the surface (roughly 2–3 weeks later), the last step is to add a 0.5-inch (1.25 cm) layer of a nutrient-poor "casing" material that has been hydrated to the same degree as the manure in step 1, and heat treated as in steps

4–8. A popular casing recipe is 1 part agricultural lime, 3 parts calcium carbonate, 12.5 parts hydrated vermiculite, and 12.5 parts hydrated peat moss. Be sure to add this "casing layer" when the mycelium is just starting to peek through the substrate top, and not after the mycelium has grown over the surface and formed a solid layer.

15. Recover the tray with the perforated foil and set it in an area where it will not be disturbed.

16. Once the mycelium has grown through the casing and is just starting to be visible (roughly 1–2 weeks after adding the casing), remove the foil and place the tray in the fruiting environment.

17. Mushrooms should start to form several days later and are ready to harvest any time before their caps start to flatten out. Enjoy!

As with any new skill, mushroom cultivation is best met with a blend of patience and perseverance. While the above techniques may feel foreign or a bit tech-geeky at first, with time and a bit of bumbling through a few inevitable moldy mistakes, these cornerstone methods will soon become invaluable life skills you'll forever be able to share with others, anywhere you go.

MUSHROOMS ON *Manure*

1 PASTEURIZE SUBSTRATE

4 PASTEURIZE CASING

2 MIX & HYDRATE SUBSTRATE

5 MIX & HYDRATE CASING

3 INOCULATE SUBSTRATE

6 APPLY CASING

3 INCUBATE

6 INCUBATE

Growing mushrooms on manure-based substrates starts with mixing, hydrating, and pasteurizing several ingredients, which are then inoculated with grain spawn in a shallow pan. After a period of incubation, a layer of pasteurized casing material is added to the pan and left to incubate for several days, after which point fruiting is initiated.

Chapter 5: THE LANGUAGE OF A MYCOFOLK

*W*hether we see little brown mushrooms on the forest floor as beautiful or basic is largely influenced by how these common fruitbodies have been presented to us by our friends, family, and society in general. As the mycoculture now evolves, all of us have the opportunity to reconsider the paradigms we've inherited on the value of fungi and replace or refine those beliefs simply through the words, symbols, and gestures we use to celebrate the human-fungal relation.

Compared to the twentieth century's limited contexts for discussing fungi (primarily in academia or industry), today's mycophiles are expressing their passion for mycology with greater diversity and creativity than ever before. Modern paintings, experimental films, digital memes, dance pieces, short stories, art installations, and illustrations are being produced every day that celebrate all forms of fungi— each with their creator's unique message from the mycelium. Alongside the research efforts of mushroom cultivators advancing our understanding of applied mycology, this creative surge is playing an essential role in ensuring that the not-too-distant future of fungi is more deeply woven into human life than we can currently imagine.

Co-creating the mycoculture's current course through creative expression is one of the revolution's most exciting aspects. How to best contribute to this growth is a question that remains open-ended for each of us, but one that is answered in short time by following the excitement and intentions drawn out by each new species that crosses our path.

MYCELIUM AS THE MEDIUM

One of the most direct ways to convey the beauty of fungi is by working with their primary tissue as a creative medium. Building on the skills covered in the previous chapter, recent advancements in cultivating fungi have led to the realization that mycelium can be grown into functional objects of nearly any shape and—depending on the species and growing conditions—with a range of unique qualities. Such mycomaterials can be made lightweight, buoyant, biodegradable, and fire-retardant, while also being quite durable and highly resistant to impacts. This novel material has rapidly been gaining in popularity around the world as a sculptural medium for artists, while also holding a high potential for pragmatically replacing plastics, pressboard, and leather in various processes—simply by using cultivation techniques that are inexpensive and easy to replicate.

The process for growing functional mycelium is fairly straightforward and, in one approach, follows the same technique used for growing wood-loving mushrooms detailed in Chapter 4. The difference here is that the container used is a plastic or metal form that provides a defined shape to the mycelium, (e.g. a gelatin dessert mold or custom 3D printed form), and which is not deeper than 10 inches (25 cm). After being inoculated, the entire substrate-filled mold is then placed in a clean plastic bag fitted with an air filter, so that the material doesn't dry out, and left to incubate for several weeks as the mycelium grows through the substrate. Once fully myceliated, the form is then removed from the bag to dry, before being popped out of the form and fully dried in the sun or in an oven set to a low temperature.

While the best results with this practice are typically had by combining actively growing grain spawn with freshly prepared sawdust, I've also had success crumbling Reishi mycelium that has already fruited, and then lightly packing that material into a clean form, while lightly misting the myceliated substrate with water as I go. If you aren't able to grow fresh mycelium, but you have access to the spent substrate from a nearby grower, this slightly more experimental approach is an excellent and easy way to play with the creative potential of mycomaterials for free.

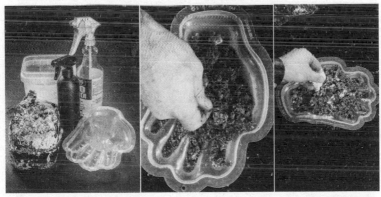

Making mycelium-based objects is easy! Crumble a bag of previously-fruited sawdust-based substrate (Reishi is an ideal species), and mix it with a light amount of potato starch and a fine mist of filtered water. Semi-firmly pack the mix into a mold, and then place this in a clean environment where it will not dry out. After the mycelium has run through, dehydrate it completely and enjoy!

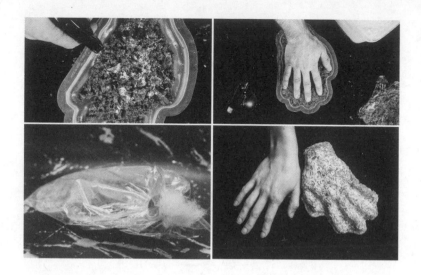

UPCYCLED MYCOPAPER

Another method for upcycling mushroom mycelium into art is by turning the fruitbody pieces left over from making medicinal mushroom tea (discussed in Chapter 3) into paper. This process is similar to that used to make paper from fibrous plants or old newspapers (skills that are easy to learn online or through books from your local library). As with these more common paper-making practices, mushroom paper is made by grinding the tissue as finely as possible, diluting the material in water, optionally mixing in other materials or binding agents, and then running the resulting soup through a screen to produce a thin sheet of wet fibers that is dried using towels, felt, or a paper-making press. A versatile, creative medium, mycopaper can be used as a canvas or even infused with spores to provide a secret inoculum to a faraway pen pal.

One of several ways to make mushroom paper. A plastic mesh is placed between a mold and deckle, which are dipped into a slurry of water and ground mushrooms, pulled up level, then flipped onto heavy paper or fabric to dry.

DYE BY FUNGI

The chemical complexity of mushrooms extends from their medicinal, psychoactive, and toxic compounds to molecules that can dye natural fibers. Simply boiling a mushroom that is known to contain dyeing pigments for an hour is all that is needed to release these coloring chemicals. Once the water is cool, cotton, wool, or even your own hair can be dipped in the water to take up the color. Some of the best mushroom species to work with as dye sources include *Gymnopilus spp.* (yellow dye), *Hydnellum spp.* (brown dye), *Hypholoma spp.* (yellow dye), *Inonotus spp.* (yellow dye), *Phaeolus schweinitzii* (green, red, or burnt sienna dye), *Phellinus spp.* (yellow dye), *Pycnoporus cinnabarinus* (orange dye), *Suillus spp.* (yellow dye), and *Thelephora spp.* (blue dye), though feel free to experiment with other non-toxic species you find in a field or forest near you!

MOVE LIKE MYCELIUM

Complementing the visual arts that express mycophilia are the physical ways we can use our bodies to explore the beauty of fungi. One example of embodied mycology is to blindfold yourself and slowly, safely, and silently walk or crawl through a natural space with only your non-visual senses to guide your interpretation of the world—much like a hyphal tip. This exercise is best done with a non-blindfolded partner who not only ensures that you don't get hurt, but who can also quietly hand you mushrooms or other natural objects affected by fungi to most fully engage your senses. This simple exercise is an easy means to begin thinking like a mushroom and to imagine navigating soils or the microscopic world beneath the bark of a fallen log.

SPREAD YOUR SPORES!

After practicing the above techniques, consider branching out and mixing your fungi love with whatever creative expression calls to you. Some of the many ways fungi can (and should!) be incorporated into our daily lives include:

- Create a fungi-themed band or solo musical act.

- Create a fictional short movie or mini documentary that involves fungi.

- Keep a mycology journal, detailing your experiences and excitement exploring the topic.

- Design a mycelium-based board game.

- Start a backyard garden of local (and abundant) lichens.

- Paint a fungi-inspired mural.

- Record music using only mushrooms and lichens as instruments.

- Make mushroom paper with an annual conk and then decorate it with spore-infused ink.

- Write a fungi-themed poem or short story.

- Give or receive a mushroom-themed tattoo.

- Create edible, medicinal, or guerrilla mushroom installations at local community gardens, food banks, or environmental organizations.

- Be an active member in (or start!) a local mycological society and document the mushroom hunting stories of all of the group's members.

- Make bread, beer, mead, wine, or cheese that incorporates edible molds or mushrooms.

- Host a mycophilic costume party, potluck, or speakeasy.

- Embroider your favorite mushroom on the back of your most cherished jacket.

The more unique your expression of these or other practices, the more elaborate our mycoculture's language becomes. As you dive deeper into the art of working with fungi, don't hold back in your creativity! Each representation of mycophilia only serves to expand our shared understanding of the human-fungal potential, while also spreading *insporation* to the next generation.

A mycoremediation-themed art installation designed by the author in Geneva, Switzerland.

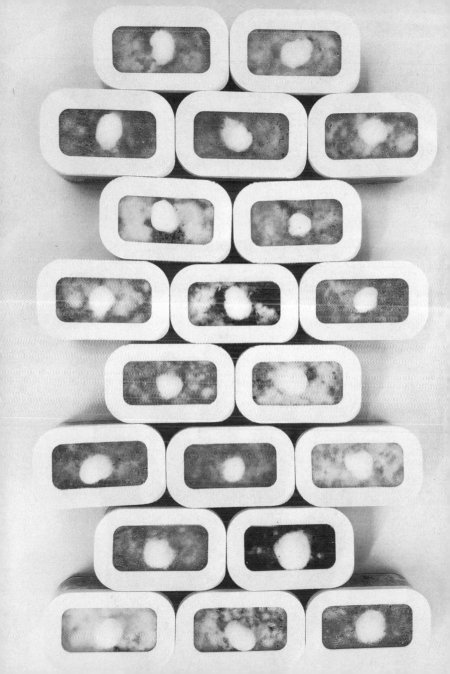

part 3

A Future for Fungi

As the mycoculture expands, so too does awareness around the potential for humans to elevate the health of our planet by working with fungi. Recent advancements in understanding the influences of macro and micro fungal species on the environment have proven that their healing effects extend from humans and plants to nearly every niche of the natural world. As we learn more about the centrality of fungi in the sustainment of life, practical mycologists are working to incorporate these latest findings into better land stewardship methods and systems that will reduce human impacts on the world, while ensuring an ever-more resilient future for coming generations.

One area of great interest is in understanding the extent to which endophytic fungi living inside of plants (discussed in the Introduction) shape ecosystems. In the wild, many of these fungi have been shown to increase heat tolerance in their plant partners, suggesting that this effect could be applied to food crops of the future, so as to alleviate the stress of rising soil and air temperatures. Some of the support these fungi provide the plant comes from the unique chemicals that they produce, many of which have also proven to offer health benefits to other plants as well as humans. Some endophytic fungi have also been shown to create diesel fuel-like compounds, while others are known to produce the medicinal compounds found in many herbs. One endophytic fungus from the Amazon was even discovered to digest a type of plastic as its primary food source! With such potent mastery of the elements of the natural world, this large and poorly studied group of fungi is at the forefront of applied mycology research for the potential it holds in helping address a range of pressing global issues.

Looking below our feet, we now recognize that the fungi permeating the soil is critical to the nourishment of life on land. As powerful decomposers, fungi kickstart the degradation of the most complex molecules of nature (most notably those in wood), helping recycle the nutrients of life throughout the food web. With their mycelial networks, mycorrhizal fungi (also discussed in the Introduction), facilitate the movement of carbon, nitrogen, phosphorus, and minerals throughout forest soils, providing nourishment where it is needed most while also feeding countless microbes and insects along the way. Through these effects, the world's mycelial networks and the compounds they produce act as the largest carbon stores on terrestrial Earth, making their abundance and diversity a key factor in moving carbon out of the atmosphere and into the ground for long-term storage.

For these reasons and many more, the historically overlooked *funga* is finally gaining consideration in environmental surveys after centuries of being overshadowed by the more visible flora and fauna. While we will never know the number of fungal species that have gone extinct due to this oversight and the past's myco-illiterate management of natural spaces, our modern appreciation for fungal functions is elevating their position out of the understory of ecological paradigms to the center of sustainable living system designs. In the coming years of the mycocultural revolution, ensuring that micro and macro fungal species increase in abundance and diversity, both inside and outside of our homes and workspaces, will be a central aspect of a healthy human-fungal-ecological relation—one of the most important topics for current and coming generations. Luckily, engaging in this work takes many forms and can be engaged with anywhere and by anyone. Guided by the fungi, we are finding new paths away from aging systems based on environmental loss and toward a more resilient future strengthened by an interconnected web of humans, hyphae, and habitats.

Chapter 6: THE MUSHROOM GARDEN

Growing a diversity of macro and micro fungal species around our homes and community gathering spaces is one of the simplest, but most essential means for bringing more fungi to the world around us. Many cultivated mushrooms thrive in garden settings, where they provide nutrient-dense food in shaded areas that most plants don't prefer. Likewise, micro fungi can be added to soil and compost systems to increase nutrient availability and reduce the growth of root rot and other plant pathogens. In bringing these and other fungi to our land tending practices, we not only receive the gifts that these species offer, but also increase their abundance—whether nearby today or faraway tomorrow—through the spores they will come to spread.

Fungal gardening is simple to engage with, and yet offers powerful insights into the importance of fungi in the environment. In the laboratory of the mushroom garden, we witness the changes in species composition that occurs as wood decomposes and fungi-rich soils age. By inoculating plant crops with mycorrhizal species, we're able to experience the dramatic increase in growth and yield that these near-invisible fungi provide to the majority of the world's plants. And through their increased presence in our daily life, we can more fully appreciate their habits and unique beauty, which complements and contrasts with flowers and foliage. The species listed below are safe to work with as they are not considered invasive, parasitic, or otherwise problematic for landscaped areas.

The mushroom garden is a place unbound by time. On one hand, it is as ancient as the first attempts to grow food crops by humans—an experimental process in which micro fungi were unknowingly a part of through their influence on soils. And on the other, it is as revolutionary as the recent advances in indoor mushroom cultivation and the many prospects that applied fungal ecology research suggests for healing the world.

Adding to this timeless practice is an essential part of ensuring that the mycoculture remains abundant in the future, despite any challenges that come its way. As we spread spores, we also spread hope for an increasingly resilient tomorrow. All that is needed to play with and learn from wild mycelium is the time and patience it takes for these networks to establish and sink deep as we prepare for the times ahead.

QUEEN STROPHARIA: THE GIVER

Just as growing Oysters mushroom on wheat straw or coffee grounds (skills covered in Chapter 4) are among the first indoor techniques beginning growers should learn, the Queen Stropharia or Garden Giant mushroom (*Stropharia rugosoannulata*) is by far the best species to begin playing with in the garden. Versatile, hearty, and nearly bulletproof, this incredible mushroom is one of the easiest and quickest to grow outdoors, and also

*The magnificent Garden Giant (**Stropharia rugosoannulata**) is a medium to large mushroom with a burgundy cap, dense white stalk, and dark colored gills. Easy to grow on a range of woody and composted materials, it is the premiere member of any mushroom garden.*

one that is tolerant of drier climates, warmer temperatures, sun exposure, and less-than-ideal (e.g. aged) substrates. In addition, its nickname, Garden Giant, is inspired by its massive fruitbodies, which can form within months of it being installed outdoors and may reach dinner plate sized proportions!

The first step to establishing a Stropharia bed is sourcing sawdust spawn for the mushroom along with fresh substrate. The ideal substrate is hardwood chips that were obtained from a healthy tree and which have been soaked in non-chlorinated water for 24 hours. If fresh wood chips aren't available, older hardwood chips, or hardwood sawdust mixed with chopped and wetted wheat straw are good alternatives.

Once the materials are gathered, installing the bed only takes about an hour, with the first step being to select a site that is in partial or full shade, regularly receives water, drains well, and

A decorative Garden Giant installation at the site of the 2016 Radical Mycology Convergence in upstate New York.

is frequently visited (so that you don't miss the mushrooms when they fruit). Decide on the shape of the mushroom bed, then dig down 6–8 inches (15–20 cm) to create a depression for holding the substrate and for water to collect in. Line the bottom of the hole with tape-free cardboard, then mix the substrate and spawn at a ratio of approximately five pounds of spawn for every 50 square feet (4.65 m²) of the bed. Water the materials lightly as you fill in the dug out area, then cover the bed with a loose layer of cardboard or 6 inches (15 cm) of fresh straw to prevent the material from drying out in the coming months.

If installed in the spring, mushrooms tend to appear after 3–6 months, and often produce several crops for 2 years if no maintenance is done. To extend the yield, buckets of myceliated

Installing a Garden Giant patch on your land is as simple as digging a shallow depression in a shaded area, lining the hole with moist cardboard, filling it with a mix of fresh hardwood chips and sawdust spawn, and then topping the materials with fresh straw.

wood chips can be used to make new mushroom beds around your home or town, with the chips removed being replaced with fresh substrate to help sustain the initial bed.

PLUGGING IN TO MUSHROOM LOGS

Dating back nearly 1,000 years, the inoculation of fresh hardwood logs with Shiitake mushrooms is one of the most ancient cultivation methods. In the centuries since first being practiced, Shiitake farmers have bred a wide variety of choice strains of the mushroom, each with a fruiting temperature preference that is warm, cold, or wide-ranging. This Shiitake strain diversity enables growers today to easily find the best variety for their climate and to match different strains to each season.

Once you have sourced an ideal Shiitake strain for your region, the next step is to find an oak, alder, or maple log that was

After selecting a freshly-cut and non-diseased hardwood log, introducing Shiitake mycelium is as easy as drilling holes around the log, filling these holes with myceliated furniture dowels, covering the openings with melted beeswax, and then labeling and incubating the logs for 12–18 months.

recently cut from a healthy tree. After that, the process is a matter of introducing the mushroom's mycelium beneath the log's bark, sealing the bark openings, and then waiting 6–18 months for the mycelium to grow through the wood. After that point, the logs are soaked for 24 hours in cold water (which triggers the mycelium to produce mushrooms), and then stood upright against a pole or fence as mushrooms develop over the coming week.

After harvest, the logs are laid to rest for at least six weeks, after which time the process can be repeated, so long as the air temperature matches the preference of the Shiitake strain you are working with. Once a log starts fruiting, it can typically be fruited 3–6 times a year over the course of 3–5 years, after which point the log can be chipped and fed to a Queen Stropharia bed!

THE LICHEN GARDEN

As discussed in Chapter 1, lichens do not grow quickly and should, for the most part, be appreciated where they are found. There are a few instances, though, where harvesting lichens may be appropriate, such as when relocating lichens from an area that is soon to be deforested to a protected area; when creating an educational lichen sanctuary; or when researchers need to study

The author leading a fungi cultivation workshop at an off-grid land project in the New Mexico desert.

the effects of changes to air quality, climate, or elevation on a lichen species.

If you wish to gather lichens for one of these purposes, the method for tending them is simple. Prior to harvest, make sure that the new habitat is similar to the original one (i.e. in degrees of shading, plant community composition, soil pH, elevation, and climate). Once the lichen has been carefully transported to the new location, place it on a similar substrate that you found it on, such as a moist and mossy slope, or in the crevices of a tree trunk. If the conditions match, the lichen should be able to survive and continue slowly growing for decades to come.

THE MANY GIFTS OF AMF

Though less commonly discussed than growing mushrooms in the garden, cultivating arbuscular mycorrhizal fungi (AMF) is a vital and accessible skill that can dramatically increase the health of soils, plants, and animals with very little cost or complexity. A small group, AMF include around 300 species and yet is thought to form beneficial root symbioses with over 90% of the world's plants. After decades of extensive research, we also know that these fungi dramatically increase the production of many crop plants by improving access to water and other nutrients (especially phosphorus), while also promoting the health of their plant partners through the protection they provide to roots against soil pathogens, including other fungi. The AMF symbiosis is one of the most ancient, with fossil evidence dating to the oldest plant fossils with roots, nearly 450 million years ago! With all of their influences, these fungi are considered the most ecologically significant of all fungi, and also ones that we can luckily bring into our gardens with ease.

Cultivating AMF is as simple as applying a commercial inoculum (per the product's instructions) to a plant's root system at the time of planting. However, many AMF products have been shown through independent testing to have a significantly lower spore load than they advertise, so an alternative approach is to cultivate local AMF using one of the low-tech methods pioneered by the Rodale Institute, an agriculture research center located in Pennsylvania. This not only cuts costs and potentially increases rates of symbiosis, but also helps increase the abundance of AMF species that have adapted to your part of the world, increasing their survival for years to come.

One of the Rodale Institute's simplest approaches is to first collect a small amount of the top four inches (10 cm) of soil from a wooded area that has not been disturbed for at least two years (though longer is better). If harvested toward the end of summer, the soil is likely to have a high AMF spore load. This soil is then taken home and mixed with nine times its volume in potting soil, and then used to cultivate deep rooting local grasses, onions, or other *Allium* plant species in elevated potting containers.

After several months, the soil will ideally be filled with AMF mycelium and spores, as well as the plant's roots, which have the fungi growing inside of them. All of this material can be harvested, dried, and saved until the following spring when it is mixed with nine times its volume in fresh potting soil to create an AMF-rich mix for your backyard or community garden.

This is the most low-tech approach to working with AMF, as well as the most experimental. For more information on AMF, including more refined means for cultivating and identifying them, see the many resources on the Rodale Institute's website, as well as the exhaustive review in *Radical Mycology: A Treatise on Seeing and Working With Fungi*.

CULTIVATING INDIGENOUS MICRO FUNGI

Another valuable means for cultivating micro fungi comes from the agriculture system known as Korean Natural Farming. Among this system's many practices is a method of collecting wild molds from a healthy forest system on cooked rice, and then mixing those fungi into a garden or farm's soil. Often referred to as Indigenous Microorganisms (IMO), the collected mix of wild species includes bacteria and other microbes alongside the more visible mycelium, all of which work together once brought into a garden's soil to improve nutrient cycling and soil structure.

The essential steps to working with IMOs are to fill a lidless cedar box with partially cooked white rice, cover the box top with a paper towel, and then place the box inside of a wire cage and on the ground of an aged, shaded habitat. In 4–5 days the rice will

Mortierella *species are commonly found in healthy soil systems around the world. But where soils have been harmed by tillage or chemical inputs, their prevalence often declines significantly.*

become covered in a diversity of molds and microbes, which are ideally mostly white and have an overall earthy and pleasant smell to them.

Back home, the moldy rice is then mixed with an equal amount of brown sugar inside of a glass jar and then topped with a couple layers of paper towels before being left to ferment for two weeks. This mix is then combined at a ratio of 1:450 with wheat bran and enough water to make the material moist, but not overly saturated. After seven days of composting at 110°F (43°C), this mix is combined with an equal amount of soil and as much water as is needed to moisten the mix, but not overly saturate it, and then left to ferment for another week.

Lastly, the material is mixed with equal parts manure and enough water to moisten it, and then left to ferment for a final week. This final boost of nitrogen from the manure will cause the blend to get hot (around 130–140°F [55–60°C]), but it should quickly cool off after that. After all of these steps, you will have a potent soil amendment that is packed with a diversity of fungi and microbes that are likely missing from your garden's soil, but which play a crucial role in the health of the lands around you.

EXPERIMENT ON!

Incorporating fungi into all of the ways that humans care for their lands, crops, herds, and families will be sure to play a central part in the next chapter of our mycoculture's story. Through sharing the above techniques with those around us, we set the stage for this next great leap in understanding of how best to live on the earth, as guided by the ancient and potent fungi. The above methods are just a sample of the many ways to begin this process, as well as starting places for experimenting with novel means for increasing the abundance of micro and macro species anywhere we travel.

For example, the techniques for growing Shiitake on logs can be applied to inoculate snags and stumps in nearby forests with local varieties of Oyster mushrooms or other decomposing fungi. This will not only increase the prevalence of these species, but also provide food for wild animals and insects while supporting natural decomposition processes that build soils. Likewise, the stocks of local, place-based arbuscular mycorrhizal fungi and indigenous micro fungi that you cultivate can be shared with members of your local community to support their own plant tending practices while also spawning greater networks of the most critical fungi in the environment.

In the coming years, it is my hope that fungal cultivation will be as commonplace as herb, vegetable, and tree growing, and just as diverse in its applications. But that shift will only happen when more mycophiles shake off the uncertainty that comes with experimental approaches and instead embrace the magic and mystery of the unknown facets of working with the fungi of our lives.

Chapter 7: A MYCOCULTURAL REVOLUTION

*A*s we learn more about the ways that molds and mushrooms support the natural world, we also discover new means to make our own ways of life more sustainable. Seemingly each new insight into fungal biology or ecology discovered by researchers has the potential to be translated through the innovations of experimental and radical mycologists to tackle some of the biggest issues facing current and future generations. These steps forward are complementary, with research taking one stride that is soon followed by the leg of applied mycology. But with a relatively small number of professional and citizen mycologists currently pushing the science's many fronts forward, the limits to this potential stays defined by the number of researchers active in the lab or forest. To support the mycorevolution's spread, building a larger foundation for the science of mycology to walk on is rapidly becoming one of the most important ways to ensure that the future of human-fungal relations continues to myceliate the world.

Traditional and radical mycologists complement each other's work, with the advancements of institution-funded research being applied by grassroots experimentalists, who in turn unveil new avenues for academic inquiry.

This opportunity is a rare gift, as most other natural sciences are well established and upheld by a large number of researchers and organizations with limited need for volunteer support. Mycology, though, is a young science that remains poorly supported by academia and is in great need of non-professionals to add their own voice and findings to our shared understanding. By taking part in this more technical side of working with fungi, the mycorevolutionary researchers of today shape the local, national, and international policies and practices of tomorrow into ones that are more fungi-informed and holistic in practice. Among the many gifts of working with fungi, this bright opportunity remains far from dulled, but rather is just beginning to shine on a brilliant revolution.

DOCUMENTING FUNGI, FOR THE FUTURE

In the ancient art of hunting mushrooms we find that one of the easiest means to advance modern mycology is through the documentation of species encountered on forays via mushroom identification websites, such as Mushroom Observer (www.mushroomobserver.org) or iNaturalist (www.inaturalist.org). This important work—which can only be crowdsourced due to the amount of data needed—not only helps refine our global understanding of mushroom distribution and fruiting patterns, but also clarifies which mushrooms are threatened or endangered based on their rate of occurrence. The distribution maps this documentation will produce in time can then be used to create better protection policies for rare fungi as well as for the habitats in which they are found. Additionally, a better understanding of species distribution patterns can help ecologists assess larger environmental processes, such as wood decomposition, which in turn supports the development of land management practices that more fully account for the impacts of fungi on soils, the climate, and the diversity of wild plants and animals.

Foragers can also support their local mushroom growers and applied mycology researchers by sourcing new species and strains for cultivation experiments. As each wild mushroom strain hosts a unique set of traits and benefits, collecting novel strains is greatly needed to advance many areas of applied mycology research—such as medicine production or the breakdown of toxic chemicals (discussed below). At the same time, sourcing these place-based strains encourages the cultivation of local foods and the creation of sustainable and circular economies, wherein fungi adapted to a region's climate are sustainably grown on the agricultural waste produced by nearby plant farmers, with the remains thereafter used to feed compost piles and livestock.

A small portion of the extensive fungarium collection at the Royal Botanical Gardens in Melbourne, Australia.

Conservation-minded mushroom hunters can also collect the spores and tissue cultures of wild fungi to preserve the genetics of endemic species and strains, much in the same way that plant growers preserve and protect the seeds of heirloom varietals and pass them on to future generations. Mushrooms and lichens collected on forays can likewise be submitted to universities to enrich their fungaria. This physical record provides a wealth of genetic and visual information for

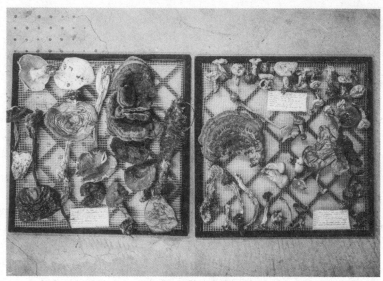

Dehydrated wild harvested mushrooms ready for adding to the MYCOLOGOS fungarium.

future mycologists to learn from that can't be provided through notes or photographs.

For me, all of these applications of mushroom hunting are equally important, and in combination demonstrate how your personal interest in collecting mushrooms has layers of potential impacts on the future (just as all of our actions do!). Whenever I go out foraging I not only look for mushrooms that I can eat or make medicine from, but also new species to photograph and document, cultivate or collect spore prints from, or simply appreciate for their ecological contributions as I keep a keen eye out for the chance to discover and name a species new to science.

MYCOREMEDIATION: APPLIED MYCOLOGY FOR THE PLANET

I believe that one of the most important areas of applied mycology research is working with fungi to break down human-developed pollutants. This process, known as mycoremediation, has been heavily researched since the 1980s by mycologists and chemists around the world. Today, we know that many molds, yeasts, and pan-niche mushrooms have the natural ability to degrade a range of toxins with ease—from plastics to pesticides to PCBs. The methods involved are low-cost, low tech, and quite similar to those discussed in Chapter 4 to grow mushrooms. And yet, mycoremediation research has rarely been scaled up from the lab bench to field-based installations that could best prove its potential to address real-world pollution problems.

Much of this delay has been due to the combined factors of 1) mycoremediation being so new that most government and environmental agencies aren't aware of—or interested in—its potential to replace traditional remediation strategies (such as burning polluted soil), and 2) that fungal cultivation and the other skills needed to efficiently practice mycoremediation have historically been inaccessible to most people, limiting the potential number of citizen researchers.

These two hurdles are why I have always promoted a grassroots, frontline approach to the advancement of mycoremediation as a science developed by and for the people most affected by environmental degradation. Rather than wait for bureaucracies to slowly determine the practice's value, dedicated individuals and groups could rapidly expand the science once they are comfortable with the fundamental skills and concepts involved in the practice. These topics, such as experimental design, are beyond the scope of this book, but I believe are easy to understand with minimal training.

That said, demonstrating mycoremediation's potential to yourself or to your friends can be easily accomplished using a method I developed years ago to grow Oyster mushrooms on used cigarette filters, the most commonly polluted object in the world. Following the same principles for growing these mushrooms on coffee grounds (discussed in Chapter 4), the mycelium will rapidly grow through the filters, which are essentially small sponges filled with hundreds of smoke-created toxins that the mycelium readily digests.

Though any mushrooms that fruit from this experiment would not be safe to eat (nor should any of the materials be added to your compost pile), this mini myco-digester is a great conversation starter for friends and family unfamiliar with the transformative powers of fungi.

To try this at home, all that is needed are used cigarette filters that are not water-logged from rain and Oyster spawn of some sort. Grain spawn is ideal as the grains provide nutrients that help the mycelium eat the pollutants. Alternatively, you can experimentally use coffee grounds that are grown through with mycelium, or even try tearing up a fresh Oyster from the store or forest in hopes that its internal mycelium will grow on the filters.

Lightly mist the filters with water to make them a bit moist, but not dripping with water, and then layer them with the spawn

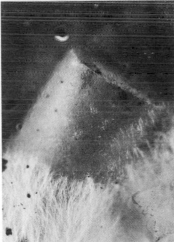

Oyster mushrooms readily consume used cigarette filters. Just combine lightly moistened filters with grain spawn (and coffee grounds, if you like, as is done here), inside of a clean jar.

inside of a clean jar. Optionally, you can add a light sprinkling of hardwood sawdust or coffee grounds to provide additional food, but the experiment is most impressive if the majority of the container is filled with hydrated filters. If the conditions are right, the mycelium will grow through all of the material, digesting many of the toxins as it does so. If grown into a shape (following the principles covered in Chapter 5), this process can even be used to upcycle an otherwise invaluable pollutant into a functional object, such as a mycoashtray!

THE MEETING OF MYCOCULTURAL MINDS

Many of the skills and concepts discussed throughout *The Mycocultural Revolution* are also in great need of research by citizen mycologists. Refining the integration of micro and macro fungi in plant cultivation practices, improving the simplicity of mushroom growing, and increasing the potency of home-made mushroom medicines are just a few of the most accessible topics in need of innovation—all of which can be moved forward with little more than curiosity, simple tools, and the patience to learn by doing.

Though mycology has only slowly gained public attention over the last few decades, these and other advancements are what is now needed to accelerate the mycocultural revolution's groundswell and bring the many gifts of mycology to people everywhere. By reflecting the great lesson of symbiosis and mutual aid that fungi express, the global community of mycophiles will be most effective in bringing this potential to fruition—both in the alliances we build with our friends and neighbors, and through increased collaboration between researchers and applied mycologists. These hyphal bridges will be the ties that draw the mycoculture's many branches closer and strengthen our combined resilience through the exchange of information and experiences.

We live in an unprecedented era in the human-fungal story: a precious and precarious time that is rapidly determining how the next chapters and volumes in this curious saga unfold. Foragers, cultivators, fermenters, artists, and all mycorevolutionaries stand at the leading edge of this growth into the unknown depths of the mycelial web's potential—each of us a raised hyphal fist proclaiming the right to shape and steward the world we share. The new future of fungi is only just beginning. And like all great revolutions of thought and culture, its winding path is unclear, but woven with the thrill of cultivating a better tomorrow and lined by a collective will to overcome adversity. In countless ways, large and small, the revolution *is* happening, all around us. How will *you* join us in the fungi?

Appendix A: SPECIES PROFILES

*T*he following seven mushrooms are considered fool-proof by most foragers as they have distinct appearances, making them unlikely to be confused for an inedible or poisonous species. The descriptions below should suffice in identifying them in the wild, but it is best to pair this book with your regional mushroom identification field guide, which will provide photographs of local varieties.

CHANTERELLES (*Cantharellus* spp.)

Chanterelle mushrooms include a number of closely related edible species that vary in their dimensions or color. Prized for their rich flavor, Chanterelles are also quite nourishing as they are high in protein and vitamins A and D2. Medicinally, they have been found to regulate calcium transportation, support dry skin, and ease eye inflammation.

The following description is for *Cantharellus cibarius*, which is common in the Pacific Northwest of the U.S. Check your local field guide for differences in your local Chanterelle species.

Habitat

- Found across North America, Europe, and Southeast Asia.

- Being mycorrhizal, it is found scattered under or near trees—most often old or second growth conifer or broadleaf trees (e.g. Douglas-fir, live oak, hemlocks, and spruces).

- The fruiting window varies by location and climate. In the Southeast U.S. it often fruits June–September. In the Northeast U.S., July–August is the common window. While in the Northwest U.S., September–November is common, compared to November–February in California.

Identifying Features

- Typically a bright yellow or yellow-orange color.

- The inner tissue is solid, dry, firm, thick, and colored like the cap or paler. The tissue tends to stain dark yellow after handling.

- There are no true gills but shallow blunt ridges that run down the upper portion of the stalk, often forking.

- Its silhouette changes from a curved to flat to an upturned shape as it ages.

- The cap is 3–25 cm wide, with a surface that is smooth or cracked and not slimy.

- The stalk is 2–10 cm tall by 0.5–5 cm wide, and more or less the same width throughout.

- Many have an apricot-like scent.

CHICKEN OF THE WOODS (*Laetiporus* spp.)

Chicken of the Woods is the name given to six closely related species in North America. As its name implies, this dense-tissue mushroom has a chicken-like texture, but with a flavor many place closer to crab or lobster. It causes stomach distress in some people, especially if found growing on coniferous wood or if eaten raw— caution is advised.

Medicinally, it has been found to inhibit the microbe responsible for UTIs and tropical infections, and to demonstrate potential in alleviating cystic fibrosis.

Habitat

- Found across North America and Europe.

- A decomposer, it is often found living on dead oak, plum, poplar, willow, beech, and conifer trees, where it causes a reddish-brown cubical heart rot.

- Typically found September to October as a cluster of individual caps in a shelving formation or rosette.

- Usually reappearing annually.

Identifying Features

- Easily recognized by its bright orange or yellow color, thick tissue, and shelving growth on wood.

- Caps range from 5–30 cm across, up to 20 cm deep, and up to 3 cm thick.

- Overall shape is semicircular or like a fan, with a wrinkled form and suede-like texture.

- Inner tissue is thick, soft, white or pale yellow, and watery, tough, or crumbling, depending on age.

- Pores on the underside number 2–4 per millimeter and are yellow and circular or angular in shape.

MORELS [*Morchella* spp.]

A choice edible which continues to defy controlled cultivation practices (despite decades of research), this umami-rich mushroom is a prized find by foragers of all backgrounds. But caution is advised as it harbors toxic compounds that must be cooked out (ideally in a well ventilated area). Medically, it has been shown to help with reducing phlegm, indigestion, excessive sputum, and shortness of breath.

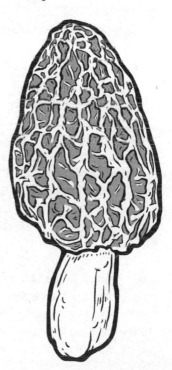

Habitat

- Found in the spring throughout North America, Europe, and Asia.

- Seemingly a decomposer and mycorrhizal mushroom, it is often found solitarily or in scattered groups in burned areas, campgrounds, old apple orchards (where it may accumulate toxic arsenic), or under hardwood or coniferous trees (e.g. white ash, green ash, elms, aspens, balsams, poplars, sycamores, tulips, pines).

Identifying Features
- Easily recognized by its honeycomb cap, hollow center, and attachment of cap to stalk.

- The stalk is 2–8 cm tall by 1–3 cm wide. Its width is typically consistent, though the base may be slightly swollen. The exterior is whitish or pale brownish, with more or less smooth texture and occasional folds.

- The cap is 3–8 cm tall by 2–5 cm wide with a conical or bluntly conical shape. The ridge color varies from tan to brown to dark brown to black, while the pits are dull brownish-yellow or olive.

SHAGGY MANE (*Coprinus comatus*)

This cosmopolitan species is found in a range of habitats around the world. A tasty species, it is best enjoyed the same day it is harvested as it quickly degrades into a pool of black ink (this can be delayed by dipping it in cold salt water or by pan frying it for a few minutes). Medicinally, it has been shown to fight breast cancer in vitro, lower blood sugar in diabetics, mitigate urinary tract infections, and alleviate lung disease. As a dye mushroom it produces a bayberry color when using an iron pot and gray-green when using an ammonia bath.

Habitat

- Found spring–early winter nearly worldwide.

- Most often in lawns, meadows, enriched soils, and gardens, and in disturbed habitats such as roadsides and along trails. Can lift 3 inches of asphalt.

Identifying Features

- Easily recognized by its shaggy appearance and tall, slender shape.

- The stalk is 5–40 cm tall by 1–2.5 cm wide, often tapering upward with a smooth texture and white color. Interior is hollow or pith stuffed.

- The cap is 4–25 cm tall, white, cylindrical or bell-shaped, and with scales. As it ages, the cap expands and curls up, eventually liquiquifying from the bottom up. Interior flesh is soft and white.

- The gills are closely crowded, not attached to the stalk, and are white, pink, pink-red, or black, depending on age.

GIANT PUFFBALL (*Calvatia gigantea*)

A trophy mushroom of meadows and fields, this large white globe of a mushroom is easy to spot many yards away. Though not particularly flavorful, it can be fried and flavored a number of ways to feed a family for many fungal feasts! Just be sure to cut the fruitbody in half to check for the outline of a young mushroom within. If you find this,

you don't have a puffball but another mushroom—potentially a lethal *Amanita*!

Habitat
- Found summer to fall in grassy areas of temperate Europe and central and eastern North America.
- Found in isolation, in groups, and occasionally in ring or circle formations.

Identifying Features
- Notable for its large size, finely velvety or smooth texture, white exterior, and white, marshmallow-like interior.
- Size typically ranges from 10–50 cm across.
- Interior consistently white when young; turning olive green in age as spores develop within.

CAULIFLOWER (*Sparassis* spp.)

A name given to several closely related North American species, Cauliflower mushrooms are easy to recognize and cherished for their large size and rich, floral flavor.

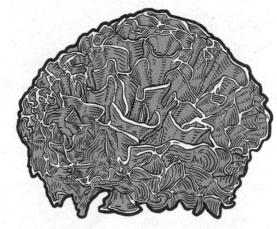

Habitat
- Found as a single mass in temperate

regions of North America and Europe in late fall to early winter.

- Grows on living or dead roots or stump bases of various conifers and hardwoods.

Identifying Features

- Recognized by its light-colored, wavy, and layered form at the base of trees.

- Size ranges from 10–60 cm across.

- Composed of tightly packed branches that arise from a shared base.

- Layers are thin and evenly colored whitish, yellowish, or tan.

HEN OF THE WOODS (*Grifola frondosa*)

An incredibly delicious mushroom, Hen of the Woods can be cultivated indoors but is *much* more delicious when found in the wild. Medicinally, it is highly regarded as  an adaptogenic species with a variety of benefits to the body—including immune support.

Habitat

- Found late summer to early fall in northern temperate deciduous forests of Japan, China, Europe, and Eastern North America.

- A root decomposer, it fruits at the base of old trees and stumps. It is most often found on oak, though it occasionally grows near chestnut, elm, maple, blackgum, beech, or larch trees.

Identifying Features

- Recognized by its large mass of wavy gray-to-brown layers at the base of trees.

- The surface is smooth, dry, and tough.

- Layers are 2–10 cm broad, and shaped like a spoon, tongue, or fan-shaped.

- The underside has 1–3 white or yellowish pores per millimeter.

Appendix B: SUPPLIERS AND ONLINE RESROUCES

MUSHROOM IDENTIFICATION RESOURCES

Matchmaker

s158336089.onlinehome.us/1an

A free, downloadable aid for identifying mushrooms.

Mushroom Expert

mushroomexpert.com

A dense site with many mushroom descriptions and tips on photography and identification.

Mushroom Observer

mushroomobserver.org

A site for amateur and professional mycologists to record observations, help others identify mushrooms, and track distribution patterns.

Mykoweb

mykoweb.com

This site includes descriptions of over 400 species and nearly 5,000 photographs.

Pacific Northwest Key Council

svims.ca/council

The Pacific Northwest Key Council is dedicated to the creation and publication of field keys to the fungi of the Pacific Northwest. Various resources are provided for free.

FERMENTATION RESOURCES

Cómo Conseguir Kéfir

lanaturaleza.es/bdkefir.htm

A Spanish site for trading tibicos, milk kefir, and kombucha cultures in the U.S. and many Spanish-speaking countries.

Cultures for Health (USA)

culturesforhealth.com

Commercial supplier of tibicos, kefir, kombucha, tempeh, and koji cultures.

GEM Cultures (USA)

gemcultures.com

Commercial supplier of tibicos, kefir, kombucha, and koji cultures.

The Kefir Shop (UK)

kefirshop.co.uk

Commercial supplier of tibicos, kefir, ginger beer plant, and kombucha cultures.

Kombucha Kamp (US)

kombuchakamp.com

Commercial supplier of kombucha cultures.

Tempeh.info

tempeh.info

Commercial supplier of tempeh cultures.

Yemoos

yemoos.com

Commercial supplier of tibicos, kefir, and ginger beer plant cultures.

Winemaking Talk
winemakingtalk.com/forum
Free discussion forum on winemaking.

Wine Press
winepress.us
Free discussion forum on winemaking.

MYCOLOGICAL ORGANIZATIONS
European Mycological Association
euromould.org
An association overseeing the advancement of mycology in Europe. Many European countries have their own mycological society that can be found online.

Fungi Foundation
ffungi.org
A non-profit working to document and protect wild fungi worldwide.

International Mycological Association
ima mycology.org
A non-profit representing 30,000 mycologists worldwide, working to further the advancement of the science.

Mycological Society of America
msafungi.org
A scientific society dedicated to advancing the various facets of mycology. Publisher of the scholarly journal Mycologia.

North American Mycological Association
namyco.org
Overarching organization for many mycological clubs and associations in North America.

Appendix C: HELPFUL CONVERSION RATIOS

VOLUME	
Imperial	Metric (Approximate)
1/8 t.	0.5 mL
1/4 t.	1 mL
1/2 t.	2.5 mL
3/4 t.	4 mL
1 t.	5 mL
1.25 t.	6 mL
1.5 t.	7.5 mL
1.75 t.	8.5 mL
2 t.	10 mL
0.05 T.	7.5 mL
1 T. (3 t./0.5 fl. oz.)	15 mL
2 T. (1 fl. oz.)	30 mL
0.25 C. (4 T.)	60 mL
0.33 C.	90 mL
0.5 C. (4 fl. oz.)	125 mL
0.66 C.	160 mL
0.75 C. (6 fl. oz.)	180 mL
1 C. (16 T./8 fl. oz.)	250 mL
1.25 C.	300 mL
1.5 C. (12 fl. oz.)	360 mL
1.66 C.	400 mL
2 C. (1 p.)	500 mL
3 C.	700 mL
4 C. (1 qt.)	950 mL
1 qt. + 1/4 C.	1 L
4 qt. (1 gal.)	3.8 L

WEIGHT	
Imperial	Metric
0.25 oz.	7 g
0.5 oz.	14 g
0.75 oz.	21.25 g
1 oz.	28.35 g
1.25 oz.	35.44 g
1.5 oz.	42.53 g
1.66 oz.	47.1 g
2 oz.	56.7 g
3 oz.	85 g
4 oz. (0.25 lb.)	113.4 g
5 oz.	141.7 g
6 oz.	170.1 g
7 oz.	198.4 g
8 oz. (0.5 lb.)	226.8 g
12 oz. (0.075 lb.)	340.2 g
16 oz. (1 lb.)	453.6 g
35.28 oz. (2.2 lbs.)	1 kg

TEMPERATURE	
Fahrenheit	Celsius
130°F	54.5°C
140°F	60°C
150°F	65.6°C
160°F	71.1°C
170°F	76.7°C
180°F	82.2°C
190°F	87.8°C
200°F	93.3°C
215°F	101.7°C
250°F	121.1°C

GLOSSARY

Adaptogen: A natural substance that helps the body adapt to stress and that exerts a normalizing effect on bodily processes.

Anastomose: To self-fuse together, as when two hyphae connect.

Arbuscule: A finely branched organ produced by glomeromycota fungi inside host root cells. The interface through which a fungus and plant exchange nutrients.

Asexual: Here used to refer to the production of spores that have not undergone genetic recombination.

Casing: A nutrient-poor water-holding top-dressing used in mushroom cultivation to increase fruitbody yields.

Deadible: Slang for a lethally toxic mushroom.

Diaspore: A self-cloning structure used by some lichens in which a small "packet" of the species' bionts bud off the main thallus and disperse. Found as *isidia* or *soredia*.

Endophytic fungi: Fungi that grow systemically within plants without causing negative symptoms.

Enzyme: A protein that speeds up specific chemical reactions.

Ergosterol: A sterol found in cell membranes of fungi, serving many of the same functions that cholesterol serves in animal cells.

Ethnomycology: The study of fungi in folklore, fiction, and ritual from prehistoric times to the modern era.

Eukaryote: Any organism whose cells contain a nucleus and other organelles enclosed within a membrane.

Exometabolites: Metabolites exuded by fungi, such as those used to digest substrates to defend against antagonistic organisms.

Fermentation: Chemical changes in organic substrates caused by the enzymes and acids of living microorganisms.

Fruitbody: A multicellular structure on which spore-producing structures (such as basidia or asci) grow.

Fungarium: A collection of dried fungal specimens.

Fungi: (*sing.* Fungus) Non-photosynthesizing eukaryotes that digest their food externally. Most produce, and live inside, a network of branched tubes (hyphae) that grow from their tip, while a small number live as individual cells.

Hypha: (*pl.* Hyphae) The primary, tube-shaped tissue of almost all fungi.

Insporation: Slang for the inspiration felt when engaging with fungi.

Lichen: A micro-ecosystem-like organism composed of symbiotic fungi, algae, bacteria, and/or a cyanobacteria.

Mold: Micro fungi often associated with the deterioration of foods or manufactured goods of organic origin.

Monotub: A self-contained mushroom incubation and fruiting environment that typically consists of several inches of inoculated substrate within a ventilated plastic storage tote.

Mushroom: The macroscopic spore-bearing fruitbody of a fungus, often (but not always) hosting a stalk and cap in various sizes, colors, and degrees of ornamentation.

Myceliation: The growth of mycelium over and through a substrate.

Mycelium: (*pl.* Mycelia) A collective network of a fungus's hyphae; the vegetative structure of a fungus.

Myco-illiteracy: The lack of mycological understanding in a person or society.

Mycoambassador: A person who increases the degree of myco-literacy in their society.

Mycobiome: The community of endosymbiotic fungi inside a plant or animal.

Mycobiont: The fungal partner in a symbiotic relationship (e.g. mycorrhiza or lichen).

Mycooulture: The combined arts, creations, and expressions of human-fungal relationships, especially when regarded as collectively achieved.

Mycofolk: Members of the modern mycoculture.

Myconoob: A person who has recently begun to learn about mycology.

Mycophile: A person who loves fungi.

Mycophilia: The love of fungi.

Mycoremediation: The application of fungi to mitigate the impacts of pollutants on an environment.

Mycorrhiza: A symbiotic relationship between a filamentous fungus and the roots of a plant.

Organelle: Any of a number of organized and specialized structures within a living cell.

Pasteurization: The application of heat to kill mesophilic organisms.

Photobiont: The photosynthesizing partner in a symbiotic relationship.

Prokaryote: A single-celled organism that lacks a membrane-bound nucleus, mitochondria, or any other membrane-bound organelle.

Psilocybin: A naturally occurring psychedelic tryptamine produced by more than 200 fungal species.

Radical Mycology: 1) A social philosophy that describes cultural phenomena through a framework inspired by the unique qualities of fungal biology and ecology. 2) A mycocentric analysis of ecological and social relationships. 3) The application of mycological research to improve planetary wellbeing.

Resilience: The capacity to recover quickly from difficulties.

Saprobe: A heterotrophic organism that derives food from dead organisms.

Spawn: 1) A material hosting mycelium and supporting its growth. 2) Such a material that is used to inoculate a fruiting substrate.

Spitzenkörper: An organelle-like structure that guides hyphal growth and is unique to fungi.

Spore: A specialized fungal reproductive structure, often single-celled and designed to be dispersed and travel away from its parent mycelium.

Substrate: 1) The food of a fungus. 2) A substance acted on by fungal digestion. 3) The material from which a fungus produces a fruitbody.

Thallus: (*pl.* Thalli) A lichen body.

Tryptamine: A monoamine alkaloid with an indole ring structure and structural similarity to the amino acid tryptophan.

Umami: The fifth flavor, responsible for the rich savoriness of animal and fungal products.

Yeast: Single-celled fungi that live in a variety of environments, often acting as decomposers or fermentation agents.

ABOUT THE AUTHOR

Peter McCoy is a world renowned applied mycology researcher and educator who has endeavored to understand and share the world of fungi with others for over 20 years. He is the author of *Radical Mycology: A Treatise on Seeing and Working with Fungi* (Chthaeus Press, 2016), the founder of the mycology advocacy organization Radical Mycology (radicalmycology.com), a co-founder of the Fungi Film Festival (fungifilmfest.com), and the founder and lead instructor at MYCOLOGOS, an applied mycology school and experimental fungi farm based in Portland, Oregon (mycologos. world). Alongside his work with fungi, Peter can be often found making art or music, pondering mysteries, or wandering the great forests of the Pacific Northwest.